U0383992

“好程序员成长”丛书

Vue.js企业开发实战

◎ 千锋教育高教产品研发部 / 编著

清华大学出版社
北 京

内容简介

Vue.js是一套用于构建用户界面的渐进式JavaScript框架，本书主要概述Vue.js的基础语法和在实际项目开发中的运用细节，遵循Vue.js自底向上逐层设计的规范。读者既可以在一个页面中单独使用，也可以将整个项目构建成单页面(SPA)应用。

本书共13章，分为3篇。基础知识篇(第1～5章)介绍Vue.js的内置指令、Vue.js实例对象中的核心选项属性，以及如何构建单页面应用；核心技术篇(第6～10章)重点讲解Vue.js重要插件的使用，在SPA应用中路由管理、状态管理、异步请求等技术的实现细节；项目实战篇(第11～13章)结合基于Vue.js的主流UI组件库，实现了以Vue.js框架为主要技术栈的Web App项目案例，便于读者快速掌握Vue.js框架在实战中的应用。本书配套案例讲解视频，帮助读者更好地理解书中的内容。

本书内容全面，讲解细致，示例丰富，适用于尚未接触过MVVM类前端框架，或者初步接触Vue.js，以及应用Vue.js开发项目的开发者阅读。

图书在版编目(CIP)数据

Vue.js企业开发实战/千锋教育高教产品研发部编著.—北京：清华大学出版社，2021.5(2022.1重印)
(“好程序员成长”丛书)
ISBN 978-7-302-57975-5

Ⅰ.①V… Ⅱ.①千… Ⅲ.①网页制作工具-程序设计 Ⅳ.①TP392.092.2

中国版本图书馆CIP数据核字(2021)第065447号

责任编辑：赵佳霓
封面设计：胡耀文
责任校对：郝美丽
责任印制：宋 林

出版发行：清华大学出版社
网　　址：http://www.tup.com.cn，http://www.wqbook.com
地　　址：北京清华大学学研大厦A座　　**邮　　编**：100084
社 总 机：010-62770175　　**邮　　购**：010-83470235
投稿与读者服务：010-62776969，c-service@tup.tsinghua.edu.cn
质量反馈：010-62772015，zhiliang@tup.tsinghua.edu.cn
课件下载：http://www.tup.com.cn,010-83470236
印 刷 者：北京富博印刷有限公司
装 订 者：北京市密云县京文制本装订厂
经　　销：全国新华书店
开　　本：185mm×260mm　　**印　　张**：16.25　　**字　　数**：396千字
版　　次：2021年7月第1版　　**印　　次**：2022年1月第2次印刷
定　　价：69.00元

产品编号：092293-01

编 委 会

序

北京千锋互联科技有限公司(简称“千锋教育”)成立于2011年1月,立足于职业教育培训领域,公司现有教育培训、高校服务、企业服务三大业务板块。教育培训业务分为大学生技能培训和职后技能培训;高校服务业务主要提供校企合作全解决方案与定制服务;企业服务业务主要为企业提供专业化综合服务。公司总部位于北京,目前已在18个城市成立分公司,现有教研讲师团队300余人。公司目前已与国内20000余家IT相关企业建立人才输送合作关系,每年培养泛IT人才近2万人,10年间累计培养超10余万泛IT人才,累计向互联网输出免费学科视频850余套,累计播放量超9500万余次。每年有数百万名学员接受千锋教育组织的技术研讨会、技术培训课、网络公开课及免费学科视频等服务。

千锋教育自成立以来一致秉承“初心至善,匠心育人”的工匠精神,打造学科课程体系和课程内容,高教产品研发部认真研读国家教育大政方针,在“三教改革”和公司的战略指导下,集公司优质资源编写高校教材,目前已经出版新一代IT技术教材50余种,积极参与高校的专业共建、课程改革项目,将优质资源输送到高校。

高校服务

锋云智慧教辅平台(www.fengyunedu.cn)是千锋教育专为中国高校打造的智慧学习云平台,依托千锋教育先进的教学资源与服务团队,可为高校师生提供全方位教辅服务,助力学科和专业建设。平台包括视频教程、原创教材、教辅平台、精品课、锋云录等专题栏目,为高校输送教材配套的课程视频、教学素材、教学案例、考试系统等教学辅助资源和工具,并为教师提供样书快递及增值服务。

锋云智慧服务QQ群

读者服务

学IT有疑问,就找“千问千知”,这是一个有问必答的IT社区,平台上的专业答疑辅导老师承诺在工作时间3h内答复你学习IT时遇到的专业问题。读者也可以通过扫描下方的二维码，关注“千问千知”微信公众号，浏览其他学习者在学习中分享的问题和收获。

“千问千知”微信公众号

资源获取

本书配套资源可添加小千 QQ 号 2133320438 或扫下方二维码获取。

小千 QQ 号

前　言

如今，科学技术与信息技术的快速发展和社会生产力变革对IT行业从业者提出了新的需求，从业者不仅要具备专业技术能力和业务实践能力，更需要培养健全的职业素质，复合型技术技能人才更受企业青睐。高校毕业生求职面临的第一道门槛是技能与经验，教科书也应紧随新一代信息技术和新职业要求的变化及时更新。

本书倡导快乐学习，实战就业，在语言描述上力求准确、通俗易懂。针对重要知识点，精心挑选企业项目案例，将理论与技能深度融合，促进隐性知识与显性知识的转化。案例讲解包含设计思路、运行效果、代码实现、技能技巧详细讲解。从动手实践的角度，帮助读者逐步掌握前沿技术，为高质量就业赋能。

在章节编排上循序渐进，在语法阐述中尽量避免使用生硬的术语和枯燥的公式，从项目开发的实际需求入手，将理论知识与实际应用相结合，促进读者学习和成长，并快速积累项目开发经验，从而在职场中拥有较高起点。

本书特点

Vue是目前最主流的前端框架之一，其简单易学、渐进式的特点很适合初学者上手入门学习。本书尽可能站在初学者的角度，用通俗易懂的语言、丰富实用的案例，循序渐进地讲解该前端技术基础知识和该框架的应用技术，逐步培养读者编程的兴趣和能力。

通过本书你将学习到以下内容。

第1章：首先介绍Vue的基本概念及安装和使用方法，然后讲解Vue的模板语法。

第2章：讲解Vue内置指令基本指令、条件渲染、列表渲染、事件处理、表单输入绑定和Class与style央视绑定。

第3章：讲解Vue实例核心选项中数据选项、DOM渲染选项、资源选项和生命周期钩子等相关内容。

第4章：通过Webpack构建Vue项目及Vue CLI脚手架工具来讲解如何使用Webpack搭建Vue工程化项目。

第5章：通过讲解组件化开发、自定义组件、组件属性校验、组件通信和插槽等内容，帮助读者进一步了解Vue组件的相关内容。

第6章：重点讲解Vue Router路由的相关概念及配置方法和导航守卫。

第7章：重点介绍Vuex状态管理的相关知识及安装和使用方法。

第8章：讲解Vue中异步请求的概念、axios的安装与使用方法，并通过实例讲解使用axios解决并发请求、拦截器、错误处理和取消请求等问题的方法。

第 9 章：讲解服务器端渲染的概念、服务器端渲染的基本用法和 Nuxt.js 框架。

第 10 章：本章将进入项目实战章节。通过介绍一款仿"京东商城"商品信息展示的电商类 App，帮助读者初步掌握通过所学基础知识进行实战应用的方法。

第 11 章：介绍一款仿"饿了么"商家页面的 App，帮助读者深入了解相关组建的进阶操作。

通过学习本书，读者可以较为充分地学习到 Vue 相关的基础概念，并通过各章节嵌入的实例更好地理解理论知识，最后通过第 11 章和第 12 章两个实战项目上手操作，将隐性知识线性化，充分了解实际应用中所需注意的事项。

致谢

本书的编写和整理工作由北京千锋互联科技有限公司高教产品研发部完成，其中主要的参与人员有吕春林、徐子惠、潘亚等。除此之外，千锋教育的 500 多名学员参与了教材的试读工作，他们站在初学者的角度对教材提出了许多宝贵的修改意见，在此一并表示衷心的感谢。

意见反馈

在本书的编写过程中，虽然力求完美，但难免有一些不足之处，欢迎各界专家和读者给予宝贵的意见，联系方式：textbook@1000phone.com。

编　者

2022 年 1 月于北京

目　　录

基础知识篇

核心技术篇

项目实战篇

本书源代码

基础知识篇

第1章 Vue 基础入门

1.1 Vue 概述

在互联网技术的蛮荒时代，前后端开发的界限还不是很清晰，所有的工作都由后端程序员完成。随着互联网的发展，前端所负责的业务逻辑不断复杂化，前端开发岗位也逐渐被分化并独立出来专注于网页的样式制作。近几年，前端工程化这个概念被广泛地提及和讨论，在互联网高速发展的今天，前端工程师可以说撑起了互联网应用的"半壁江山"。

前端开发从最初的"切图"发展到现在各式各样的前端框架，目前最流行的三大框架：Vue.js、React.js、Angular.js，其中 Vue 以其简单易学、性能优越、渐进式等特性，在三大框架中脱颖而出。截至 2019 年 3 月，Vue 在 GitHub 上的星数已经超过其他两个框架，成为三大框架中最热门的框架。

1.1.1 MVC 到 MVVM 的演化历程

在学习 Vue 之前，我们先来了解一下常见的软件设计模式。

1. MVC 模式

模型-视图-控制器(Model-View-Controller，MVC)模式是一种经典的软件设计模式。在软件技术发展之初，MVC 模式被应用于桌面应用程序中，随着 PHP、JSP 等脚本语言的诞生，MVC 模式又被应用于早期的 Web 架构中，逐渐成为 Web 1.0 时代的主流模式。

MVC 模式的特点是使用业务逻辑、模型数据、界面展示三部分相分离的方式来组织代码结构，在 MVC 设计模式中各部分的功能如下。

Model 层对要处理的业务逻辑和数据进行操作，并且接收 View 层请求的数据，然后对数据进行计算、校验、处理，最后返回最终的处理结果，整个过程不直接与用户产生交互；

View 层是用户能够看到的并且可以与之交互的客户端界面，例如桌面应用中的图形界面，以及 Web 应用中的浏览器渲染的网页等；

Controller 层相当于 Model 层和 View 层的桥梁，负责收集用户输入的数据，从对应的模型获取数据并返回给相应的视图，以此完成交互请求，使模型和视图保持一致，如图 1.1 所示。

MVC 模式实现了 Model 层和 View 层的代码分离，Model 层专注于数据管理，View 层则专注于数据显示，Controller 层在二者之间架起了一座桥梁。将业务逻辑聚集到一个部件中，在更新不同的界面或用户产生交互时，即使访问同样的数据，也会得到不同的页面呈现，由于无须重写业务逻辑的代码，而且减少了 Model 层的冗余代码，使得 Model 层和

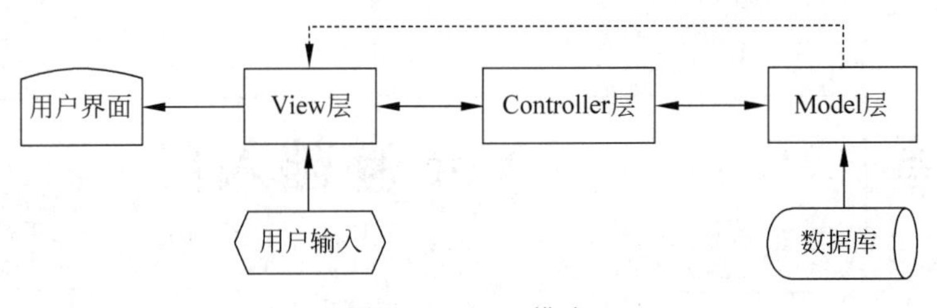

图 1.1　MVC 模式

View 层更加灵活和易于维护。由于 MVC 中 3 个模块的相互独立，改变任意一个模块都不会对另外两个模块造成影响，从而对 Model 层和 View 层进行了解耦。

伴随技术的不断迭代，MVC 模式逐渐演化出更多的形态。无论演化的是哪种版本及形态，都离不开 MVC 模式的本质。所以，在某些文章中就将这种衍生版本统称为"MV * 模式"，其中，MVVM 模式就是典型代表之一。

2. MVVM 模式

模型-视图-视图模型（Model-View-ViewModel，MVVM）模式是 MVC 的衍生版本，其主要目的是分离 View 层和 Model 层。模型指的是服务器端传递的数据，视图指的是浏览器中渲染的网页，视图模型是 MVVM 模式的核心，连接了 View 层和 Model 层。在 MVVM 模式中，主要通过两个方面实现数据的双向绑定：一方面通过数据绑定将后端传递的数据转化为用户可以看到的页面；另一方面通过 DOM 事件的监听，将用户看到的页面转换为后端数据，如图 1.2 所示。

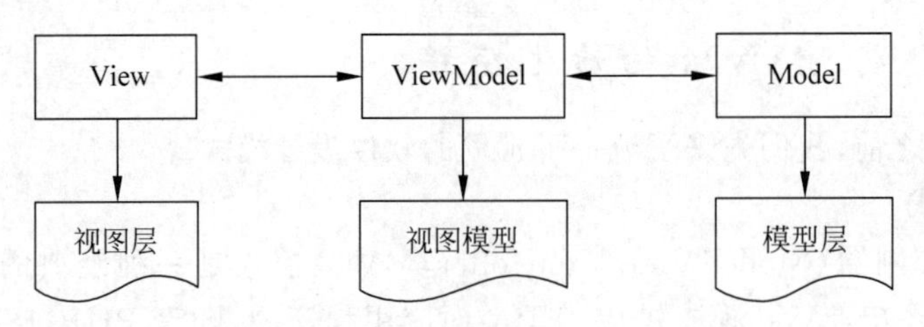

图 1.2　MVVM 模式

MVVM 模式是对 View（视图）和 Model（模型）的解耦，在接收到用户的请求后，ViewModel 获取 Model 层响应的数据，并通过数据绑定将相应的 View 层重新渲染，即 Model 层只需传入一个数据就可以实现 View 层的同步更新，从而实现 View 层和 Model 层之间的松散耦合。

Vue 就是基于 MVVM 模式设计的一套框架，在 Vue 中，JavaScript 的数据相当于 Model，例如对象、数组等，HTML 页面相当于 View，而 Vue 实例化对象相当于 ViewModel。

1.1.2　Vue 简介

Vue 是一套轻量级 MVVM 框架，与 React 和 Angular 有所不同，Vue 的核心库只关注 View 层，并且提供了简单的 API 和容易上手的操作机制，例如双向数据绑定、组件复用等，此外还很容易与第三方库进行整合。Vue 的渐进式和自底向上逐层应用的设计，使其成为十分优秀的前端框架。下面我们来解释一下什么是渐进式框架和自底向上逐层应用。

渐进式框架就是在项目开发中只需关注需要的功能特性，而不需要的部分功能可以先忽略，Vue 不强求你一次性接受并使用它的全部功能特性。

自底向上设计是一种设计程序的过程和方法，我们可以将其理解为先编写项目的基础代码部分，然后逐步扩大规模、补充和升级核心功能。

1.1.3 虚拟 DOM 与 Diff 算法

在传统的开发模式中，原生 JavaScript 操作 DOM 时，浏览器会从构建 DOM 树开始从头到尾执行一遍。例如，在一次操作中需要更新 10 个 DOM 节点，当浏览器收到第一个 DOM 更新请求后并不知道后面还有 9 个更新请求，所以会马上执行渲染流程，直到最终执行 10 次。但是在每次重新更新时，前面更新 DOM 的操作都会变成无用功，浏览器在计算 DOM 节点时白白浪费了性能，如果频繁操作，还会产生页面卡顿，从而影响用户体验。

虚拟 DOM 的设计就是为了解决浏览器性能问题，若一次操作中有 10 次更新 DOM 的动作，虚拟 DOM 不会立即操作 DOM，而是将这 10 次操作更新的 Diff 内容保存到本地的一个 JS 对象中，最终将这个 JS 对象一次性 attach 到 DOM 树上，再进行后续操作，避免大量无谓的计算量。

1.2 Vue 的安装与使用

20min

本节我们来学习一下如何安装及使用 Vue 框架。

1.2.1 直接使用<script>引入

直接使用 Vue 有两种方式，一种是使用独立的版本；另一种是使用 CDN 的方式。本书在第 1～3 章使用 Vue 的独立版本进行讲解，对于 Vue 的初学者也推荐使用这种方式入门。从第 4 章开始，使用 Vue 的脚手架创建项目。

1. 使用独立的版本

在 Vue 官网 http://cn.vuejs.org 下载最新稳定版本，然后使用 <script> 标签引入 HTML 页面中，Vue 会被注册为一个全局变量。

在官网上提供了两个版本：开发版本和生产版本，如图 1.3 所示。

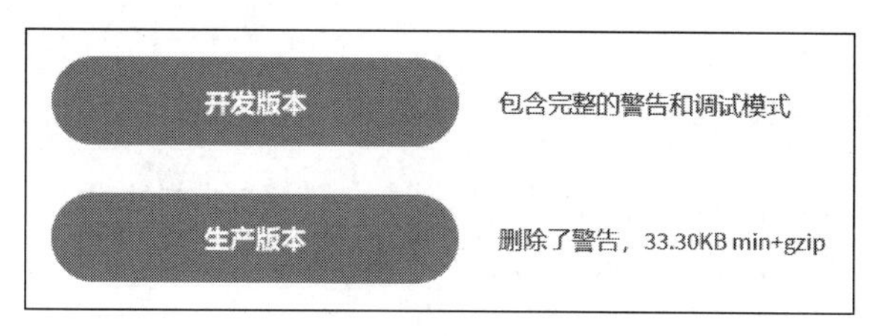

图 1.3 Vue 的下载版本

注意 在开发环境下不要使用生产版本，不然就将失去所有常见错误相关的警告！

下载完成后，直接使用 <script> 标签引入，代码如下：

```
<script src = "Vue.js"></script>
```

2. 使用CDN的方式

对于制作原型或学习，可以这样使用最新版本，代码如下：

```
<script src="https://cdn.jsdelivr.net/npm/vue/dist/Vue.js"></script>
```

对于生产环境，推荐使用一个稳定的版本号和构建文件，以避免新版本造成的不可预期的破坏，代码如下：

```
<script src="https://cdn.jsdelivr.net/npm/vue@2.6.12"></script>
```

1.2.2 使用NPM方式

在用 Vue 构建大型应用时推荐使用 NPM 安装。NPM 能很好地和诸如 Webpack 或 Browserify 模块打包器等工具配合使用。同时 Vue 也提供配套工具来开发单文件组件，安装代码如下：

```
#最新稳定版
$ npm install vue
```

由于 NPM 安装速度较慢，推荐使用淘宝镜像 CNPM，代码如下：

```
#使用淘宝镜像
$ cnpm install vue
```

1.2.3 使用命令行工具

Vue 提供了一个官方的 CLI，为单页面应用（SPA）快速搭建繁杂的脚手架。它为现代前端工作流提供了 batteries-included 的构建设置。只需几分钟就可以运行起来并带有热重载、保存时 lint 校验，以及生产环境可用的构建版本。

CLI 工具假定用户对 Node.js 和相关构建工具有一定程度的了解。如果你是新手，建议先熟悉 Vue 本身之后再使用 CLI。本书在第 4 章将详细介绍脚手架的安装及如何创建 Vue 项目。

1.2.4 创建一个Vue实例

在本节使用 Vue 独立版本，首先将 Vue.js 文件下载到本地项目目录中，在 HTML 网页文件中引入 Vue 框架，并在 <body> 底部使用 new Vue() 的方式创建一个 Vue 实例对象。

index.html 文件代码如下：

```
<!DOCTYPE html>
<html lang="zh">
<head>
    <meta charset="UTF-8">
```

```
    <title></title>
    <!-- 引入 Vue 框架 -->
    <script src="js/Vue.js"></script>
</head>
<body>
    <div id="app">
        <p>姓名:{{name}}</p>
        <p>年龄:{{age}}</p>
    </div>
    <script type="text/JavaScript">
        //创建 Vue 实例
        new Vue({
            el: '#app',
            data: {
                name: '王小明',
                age: 20
            }
        })
    </script>
</body>
</html>
```

在浏览器中打开网页,推荐使用 Google Chrome 浏览器,运行效果如图 1.4 所示。

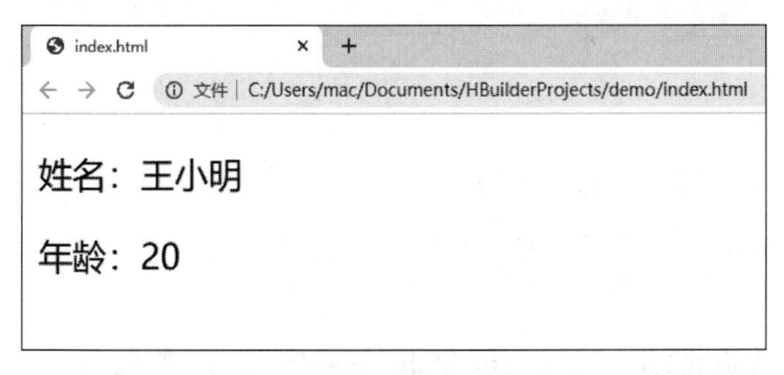

图 1.4 在浏览器中运行的效果

1.3 Vue 模板语法

Vue.js 使用了基于 HTML 的模板语法,允许开发者声明式地将 DOM 绑定至底层 Vue 实例的数据上。所有 Vue.js 的模板都是合法的 HTML,所以能被遵循规范的浏览器和 HTML 解析器解析。

在底层的实现上,Vue 将模板编译成虚拟 DOM 渲染函数。结合响应系统,Vue 能够智能地计算出最少需要重新渲染多少组件,并把 DOM 操作次数减到最少。

1.3.1 插值

插值的语法有以下 3 种:

1. 文本

数据绑定最常见的形式就是使用 Mustache 语法（双大括号）的文本插值，代码如下：

```
<span>Message: {{ msg }}</span>
```

Mustache 标签将会被替代为对应数据对象上 property 的值。无论何时，只要绑定的数据对象上 property 发生了改变，插值处的内容都会更新。

通过使用 v-once 指令，也能一次性地插值，但当数据改变时，插值处的内容却不会更新。需要留心，这会影响到该节点上的其他数据绑定，代码如下：

```
<span v-once>这个将不会改变: {{ msg }}</span>
```

在下面的示例中，向标题动态插入值，插值的内容可以根据需要进行修改，代码如下：

```
<div id="app">
    <h2>欢迎大家学习{{message}}这本书</h2>
</div>
<script>
    new Vue({
        el:"#app",
        data:{
            message:"《Vue.js 企业开发实战》"
        }
    })
</script>
```

在浏览器中运行的效果如图 1.5 所示。

图 1.5　文本渲染效果

2. 原始 HTML

双大括号会将数据解释为普通文本，而非 HTML 代码。为了输出真正的 HTML，我们需要使用 v-html 指令。

如果想要在页面中输出一个超链接，我们先在 data 属性中声明一个值，其含有<a>标签的字符串属性，然后使用 v-html 指令绑定到对应的 HTML 元素上，代码如下：

```
<div id = "app">
    <p>{{message}}</p>
    <p v-html = "message"></p>
</div>
<script>
    new Vue({
        el: '#app',
        data: {
            message: '<a href = "http://cn.vuejs.org"> Vue.js 官网</a>'
        }
    })
</script>
```

在浏览器中运行的效果如图 1.6 所示。

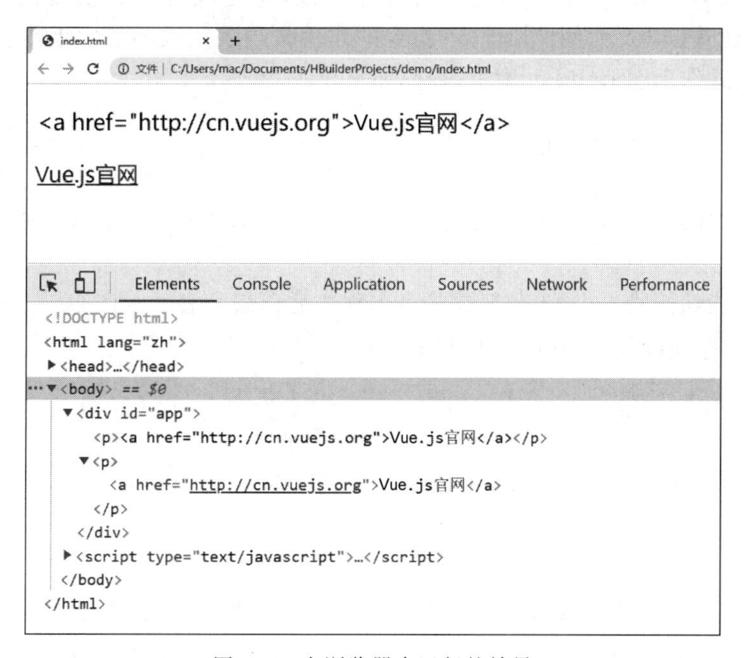

图 1.6　在浏览器中运行的效果

在浏览器的控制台中可以看到，使用 v-html 指令的 <p> 标签输出了 <a> 标签，并且在单击页面中的"Vue.js 官网"链接后，浏览器页面跳转到了对应的页面。

3. 使用 JavaScript 表达式

在前面的 Vue.js 示例代码中，一直只绑定简单的 property 键值，但实际上，对于所有的数据绑定，Vue.js 都提供了完全的 JavaScript 表达式支持，代码如下：

```
{{ number + 1 }}

{{ ok ? 'YES' : 'NO' }}

{{ message.split('').reverse().join('') }}

<div v-bind:id="'list-' + id"></div>
```

这些表达式会在所属 Vue 实例的数据作用域下作为 JavaScript 被解析。有个限制需要注意,每个绑定都只能包含单个表达式,所以下面的例子不会生效,代码如下:

```
<!-- 这是语句,不是表达式 -->
{{ var a = 1 }}

<!-- 流控制也不会生效,请使用三元表达式 -->
{{ if (ok) { return message } }}
```

1.3.2 指令

指令(directives)是带有“v-”前缀的特殊属性,指令设计的初衷是用于表示单个的 JavaScript 表达式(v-for 为例外情况)。指令的作用是当表达式的值发生改变时,可以动态地将结果响应式作用在 DOM 元素上,代码如下:

```
<p v-if="seen">这里是要显示的内容</p>
```

在上面的代码中,v-if 指令将根据表达式 seen 的值的真假动态插入或移除<p>元素。

1. 参数

一些指令能够接收一个“参数”,在指令名称之后以冒号表示。例如,v-bind 指令可以用于响应式地更新 HTML attribute,代码如下:

```
<a v-bind:href="URL">...</a>
```

这里的 href 是参数,告知 v-bind 指令将该元素的 href attribute 与表达式 URL 的值绑定。

2. 修饰符

修饰符(modifier)是由半角句号“.”指明的特殊后缀,用于指出一个指令应该以特殊方式绑定。例如,.prevent 修饰符告诉 v-on 指令对于触发的事件调用 event.preventDefault(),代码如下:

```
<form v-on:submit.prevent="onSubmit">...</form>
```

1.3.3 缩写

“v-”前缀作为一种视觉提示,用来识别模板中 Vue 特定的属性。在使用 Vue.js 为现有标签添加动态行为时，v-前缀很有帮助。然而，对于一些频繁用到的指令，这样使用会令人觉得烦琐。同时，在构建由 Vue 管理所有模板的单页面应用程序（Single Page Application，SPA）时，v-前缀也变得没那么重要了。因此，Vue 为 v-bind 和 v-on 这两个最常用的指令提供了特定简写。

1. v-bind 缩写

代码如下：

```
<!-- 完整语法 -->
<a v-bind:href="URL">...</a>

<!-- 缩写 -->
<a :href="URL">...</a>

<!-- 动态参数的缩写 (2.6.0+) -->
<a :[key]="URL"> ... </a>
```

2. v-on 缩写

代码如下：

```
<!-- 完整语法 -->
<a v-on:click="doSomething">...</a>

<!-- 缩写 -->
<a @click="doSomething">...</a>

<!-- 动态参数的缩写 (2.6.0+) -->
<a @[event]="doSomething"> ... </a>
```

它们看起来可能与普通的 HTML 略有不同，但“:”与“@”对于属性名来说都是合法字符，在所有支持 Vue 的浏览器都能被正确地解析。而且，它们不会出现在最终渲染的标记中。

第2章 Vue内置指令

31min

2.1 基本指令

2.1.1 v-text 与 v-html

v-text 与 v-html 指令都可以用来更新页面元素的内容，但是二者也有不同点，v-text 输出的数据是以字符串形式显示的，而 v-html 不仅可以输出字符串形式的数据，而且可以渲染字符串中的 HTML 标签。

index.html 文件代码如下：

```
<div id = "app">
    <p>{{message}}</p>
    <p v-text = "message"></p>
    <p v-text = "htmlStr"></p>
    <p v-html = "htmlStr"></p>
</div>
<script>
    new Vue({
        el: "#app",
        data: {
            message: "hello world",
            htmlStr: '<h2>Hello Vue.js!</h2>'
        }
    })
</script>
```

在浏览器中运行的效果如图 2.1 所示。

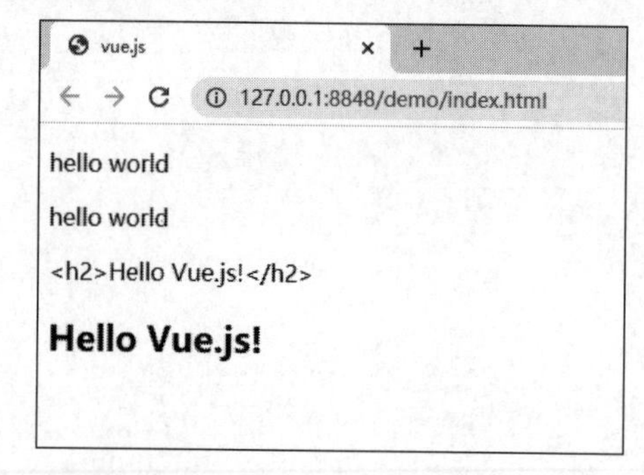

图 2.1 v-text 与 v-html 指令运行效果

2.1.2 v-bind

v-bind 可以用来绑定标签的属性，例如：< img >标签的 srcs 属性，< a >标签的 title 属性等。使用 v-bind 绑定的属性值应该为一个 JavaScript 的变量，或者是 JavaScript 表达式。

index. html 文件代码如下：

```
<div id = "app">
    <a href = "https://cn.vuejs.org/" v - bind:title = "titleText"> Vue.js 官网</a>
</div>
<script>
    new Vue({
        el: "#app",
        data: {
            titleText: "Vue.js 是目前最流行的前端框架之一"
        }
    })
</script>
```

在浏览器中运行的效果如图 2.2 所示。

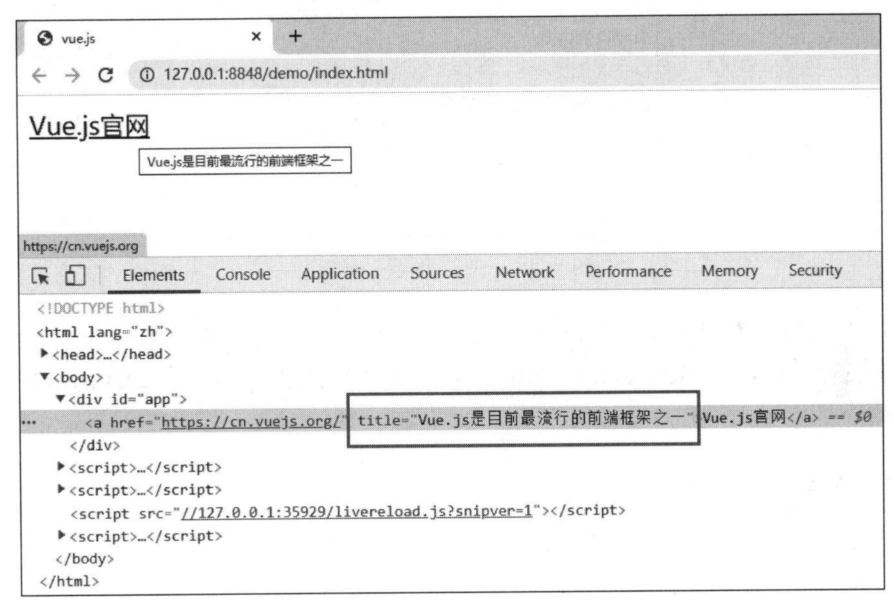

图 2.2 v-bind 指令运行效果

上面的代码示例中，对 < a > 标签的 title 属性使用 v-bind 进行绑定，title 的值为 Vue 实例中的 titleText 变量，当鼠标指针悬停在 < a > 标签渲染的元素上时，会显示动态的 titleText 属性的值。

2.1.3 v-once

v-once 指令只渲染元素和组件一次，随后的渲染，如果使用了此指令的元素、组件及其所有子节点，则都会当作静态内容并跳过。这个特性可以用于优化更新性能。

index.html 文件代码如下：

```
<div id = "app">
    <p><input type = "type" v-model = "message" /></p>
    <p v-once>只渲染一次：{{message}}</p>
    <p>可以改变：{{message}}</p>
</div>
<script>
    new Vue({
        el: "#app",
        data: {
            message: "hello"
        }
    })
</script>
```

在浏览器中运行的效果如图 2.3 所示。

图 2.3　v-once 指令运行效果

运行上面的示例代码，当打开浏览器页面时，DOM 元素中的{{message}}渲染了 message 属性的值，当使用 v-model 绑定的输入框再次改变 message 属性的值时，添加了 v-once 指令的标签则没有发生任何变化。

2.1.4　v-cloak

v-cloak 指令将会保持在 DOM 元素上，直到关联实例结束编译后自动移除。v-cloak 指令的使用场景非常有限，常用于解决网络较慢时数据的加载问题。用户在访问 Vue.js 实现的网站时，如果网络延迟，网页还在加载 Vue.js，便会导致 Vue 来不及渲染，这时页面就会显示出 Vue 源代码，我们就可以使用 v-cloak 指令来解决这一问题。

index.html 文件代码如下：

```
<div id = "app">
    {{message}}
</div>
<script>
```

```
    new Vue({
        el: "#app",
        data: {
            message: "hello world"
        }
    })
</script>
```

当网络较慢时，在浏览器中打开页面，此时会在页面中出现 Vue 源码，直到 Vue.js 加载完成并编译之后，才会显示正常的数据内容，如图 2.4 所示。

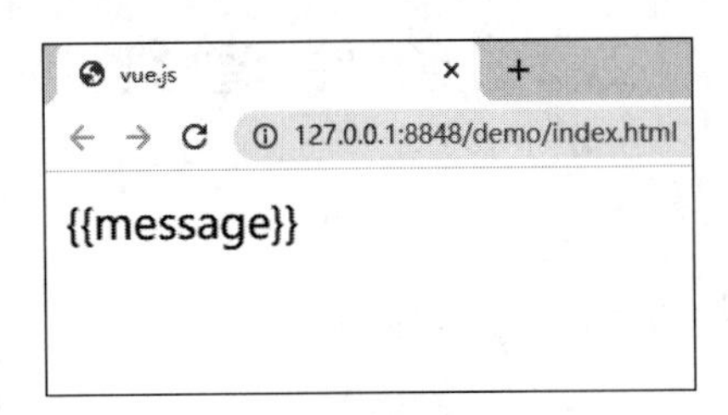

图 2.4　Vue 未完成编译的页面显示效果

为了解决上面代码运行的问题，我们可以为 DOM 元素添加 v-cloak 指令，并使用属性选择器为指定的 DOM 元素设置隐藏样式。v-cloak 指令可以保持在 DOM 元素上，直到编译结束才自动移除，这样就可以在 Vue 完成编译之前使指定的 DOM 元素处于隐藏状态，等编译完成后，CSS 样式就对该 DOM 元素无效了。

index.html 文件代码如下：

```
<style>
    [v-cloak] {
        display: none;
    }
</style>

<div id="app" v-cloak>
    {{message}}
</div>
```

2.1.5　v-pre

v-pre 指令会跳过这个元素和它的子元素的编译过程，所以可以用来显示原始 Mustache 标签。跳过大量没有指令的节点会加快编译，代码如下：

```
<span v-pre>{{ this will not be compiled }}</span>
```

在浏览器中运行的效果如图 2.5 所示。

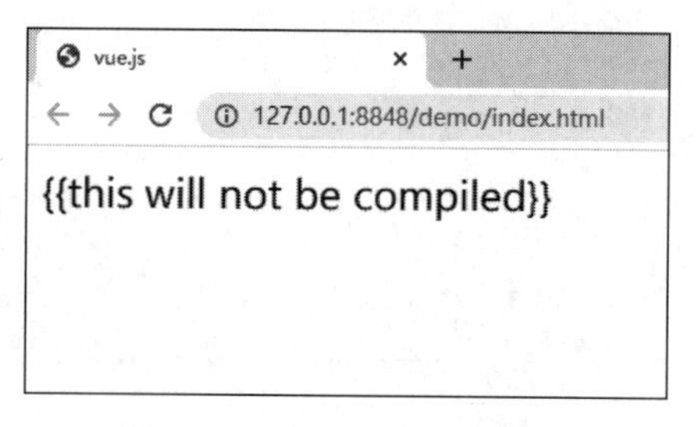

图 2.5 v-pre 指令运行效果

25min

2.2 条件渲染

2.2.1 v-show

v-show 指令可以根据条件展示元素,代码如下:

```
<h1 v-show="true">Hello!</h1>
```

带有 v-show 的元素始终会被渲染并保留在 DOM 中,v-show 只是通过简单地切换元素 CSS 属性 display:none 实现的。例如,将上面的代码 v-show 的值设置为 false,代码如下:

```
<div id="app" v-pre>
    <h1 v-show="true">Hello!</h1>
    <h1 v-show="false">World!</h1>
</div>
```

在浏览器中运行的效果如图 2.6 所示。

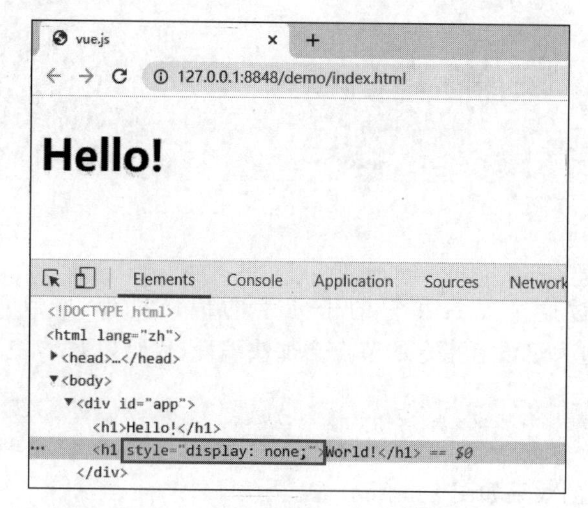

图 2.6 v-show 在浏览器中运行的效果

2.2.2 v-if 与 v-else-if

v-if 和 v-show 都可以实现条件渲染，但是 v-if 与 v-show 不同的是，v-if 不是通过切换 CSS 属性实现显示与隐藏的，当 v-if 的值为 false 时，带有 v-if 的 DOM 元素就不会被渲染出来。v-if 和 v-show 最大的不同就是，v-if 不仅可以单独使用，还可以和 v-else-if、v-else 指令配合使用，类似于 JavaScript 中的 if-else、if-else-if 语句。

1. v-if

v-if 用于条件性地渲染一部分内容。这部分内容只会在指令的表达式返回值为 true 的时候才被渲染。

index.html 文件代码如下：

```
<div id="app">
    <h2 v-if="ok">Vue.js</h2>
    <h2 v-if="!ok">React.js</h2>
</div>
<script>
    new Vue({
        el: "#app",
        data: {
            ok: true
        }
    })
</script>
```

在浏览器中运行的效果如图 2.7 所示。

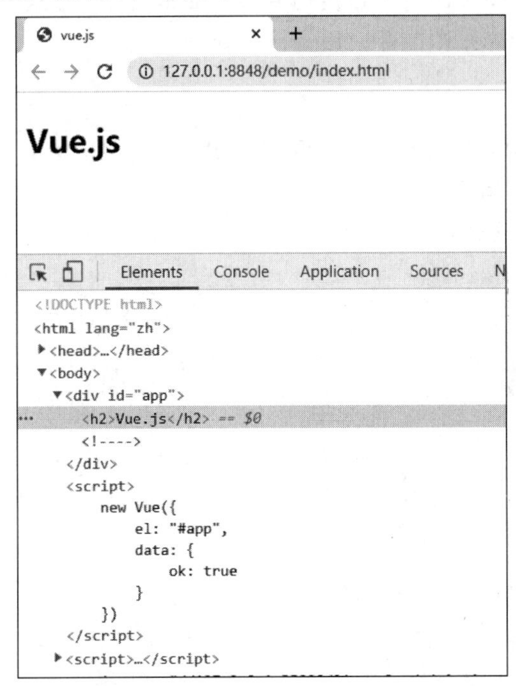

图 2.7 v-if 在浏览器中运行的效果

2. v-else-if

v-else-if 指令类似于条件语句中的“else-if 块”,可以与 v-if 配合使用。

index.html 文件代码如下：

```
<div id="app">
    <p v-if="value === 'A'">A</p>
    <p v-else-if="value === 'B'">B</p>
</div>
<script>
    new Vue({
        el: "#app",
        data: {
            value: "B"
        }
    })
</script>
```

上面代码在浏览器中运行的结果为 B,这里需要注意的是,v-else-if 指令不能单独使用,必须跟在带有 v-if 或 v-else-if 的元素之后。

3. v-if 和 v-show 的区别

v-if 与 v-show 指令都可以根据表达式的值来控制元素的显示与隐藏状态。

v-if 相比于 v-show,更“真实”地实现了元素的渲染与移除,但是在频繁的切换过程中,使 DOM 元素不断地在内存中重建与销毁,这样便增加了内存的开销。

v-show 是更简单的一种切换显示与隐藏状态的操作,只是修改了 CSS 属性中 display 的值,无论显示与隐藏,DOM 元素始终被渲染。

综上所述,如果需要在页面中频繁地切换某个元素的显示状态,推荐使用 v-show 指令;如果在运行时条件很少改变,则推荐使用 v-if 指令。

2.2.3 v-else

在根据条件渲染 DOM 元素时,还可以使用 v-else 指令来表示“else 块”,类似于 JavaScript 中的 if-else 逻辑语句。

index.html 文件代码如下：

```
<div id="app">
    <p v-if="value === 'A'">A</p>
    <p v-else-if="value === 'B'">B</p>
    <p v-else>Not A/B</p>
</div>
<script>
    new Vue({
        el: "#app",
        data: {
            value: "C"
        }
    })
</script>
```

在浏览器中运行的效果如图 2.8 所示。

图 2.8　v-else 指令在浏览器中运行的效果

2.2.4　在<template>元素上使用 v-if 条件渲染分组

因为 v-if 是一个指令，所以必须将它添加到一个元素上。但是如果想切换多个元素应该如何操作呢？此时可以把一个 <template> 元素当作不可见的包裹元素，并在上面使用 v-if。最终的渲染结果将不包含 <template> 元素。

index.html 文件代码如下：

```
<div id = "app">
    <template v - if = "value">
        <h2 > Vue.js 教程</h2 >
        <div > Vue.js 是最流行的前端框架之一</div >
        <div > Vue.js 是一套渐进式前端框架</div >
    </template >
</div >
<script >
    new Vue({
        el: "#app",
        data: {
            value: true
        }
    })
</script ></script >
```

在浏览器中运行的效果如图 2.9 所示。

图 2.9　在<template>元素上使用 v-if 条件运行的效果

2.2.5　用 key 管理可复用的元素

Vue 会尽可能高效地渲染元素，通常会复用已有元素而不是从头开始渲染。这样做除了使 Vue 的执行速度变得非常快之外，还有其他一些作用。例如，允许用户在不同的登录方式之间切换。

index.html 文件代码如下：

```
<div id="app">
    <template v-if="type === 'username'">
        用户名: <input type="text" placeholder="请输入用户名">
    </template>
    <template v-else>
        邮箱: <input type="text" placeholder="请输入邮箱">
    </template>
    <button @click="toggle">切换</button>
</div>
<script>
    new Vue({
        el: "#app",
        data: {
            type: 'username'
        },
```

```
        methods: {
            toggle(){
              this.type = this.type === 'username'? 'email' : 'username';
            }
        }
    })
</script>
```

在浏览器中运行，首先在输入框中输入“张三”，如图 2.10 所示，然后单击“切换”按钮，可以看到邮箱中显示的是“张三”，如图 2.11 所示。

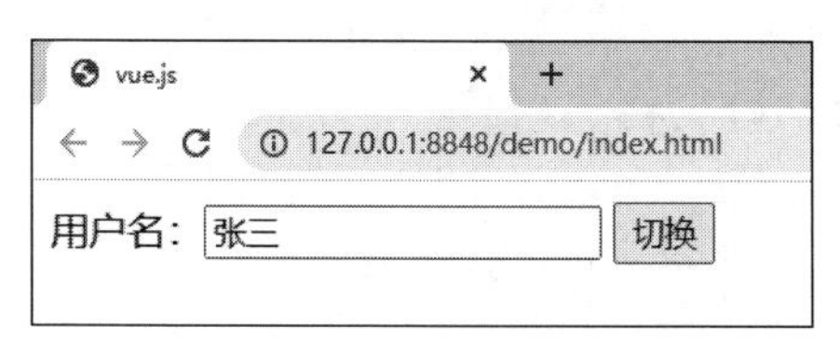

图 2.10　输入“张三”

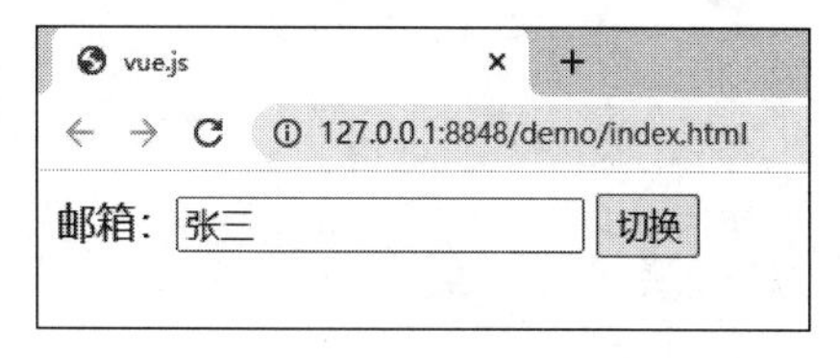

图 2.11　切换效果

在上面的示例中，切换状态并不会清空用户已经输入的内容，因为两个模板使用了相同的元素，<input>不会被替换掉，只是替换了它的 placeholder 属性。

这样也不总是符合实际需求，所以 Vue 提供了一种方式来表达“这两个元素是完全独立的，不要复用它们”。只需添加一个具有唯一值的 key 属性。

index.html 文件代码如下：

```
<div id="app">
    <template v-if="type === 'username'">
        用户名：<input type="text" placeholder="请输入用户名" key="username">
    </template>
    <template v-else>
        邮箱：<input type="text" placeholder="请输入邮箱" key="email">
    </template>
    <button @click="toggle">切换</button>
</div>
<script>
    new Vue({
        el: "#app",
        data: {
            type: 'username'
        },
        methods: {
            toggle(){
              this.type = this.type === 'username'? 'email' : 'username';
            }
        }
    })
</script>
```

在浏览器中运行，首先在输入框中输入“张三”，如图 2.12 所示，然后单击“切换”按钮，可以看到邮箱中的值为默认的 placeholder 属性的值，如图 2.13 所示。

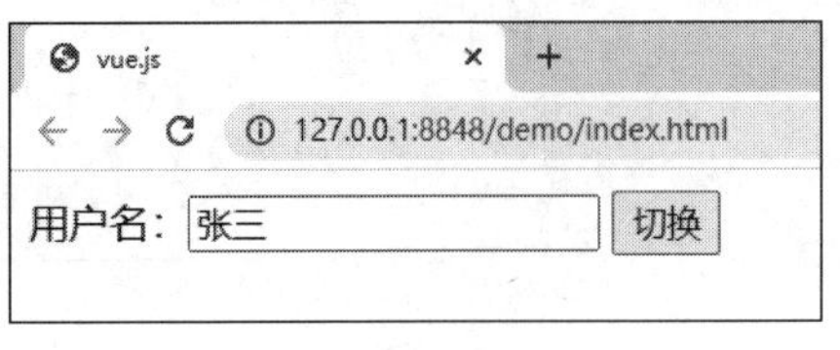

图 2.12 输入“张三”

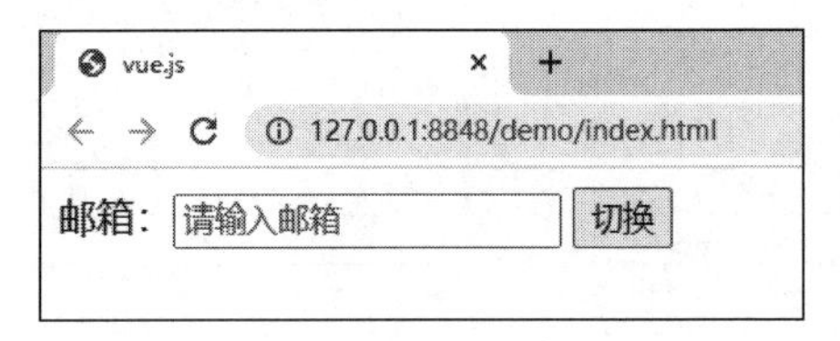

图 2.13 切换效果

2.3 列表渲染

2.3.1 遍历元素

当遍历一个数组或枚举一个对象进行迭代循环展示时，会用到列表渲染的指令 v-for。v-for 指令类似于 JavaScript 中的 for 循环，在 Vue 中提供了 v-for 指令用来循环数组。

在使用 v-for 指令时，可以对数组、对象、数字、字符串进行循环，获取其每个元素。在使用 v-for 指令时，要按照特定的 for-in 语法进行遍历，代码如下：

```
<div v-for="(item,index) in items">
    {{index}} - {{item}}
</div>
```

在上面的示例代码中，items 是源数据数组，item 为每次迭代遍历的数组元素，index 为元素在数组中的索引。

1. 遍历数组

index.html 文件代码如下：

```
<div id="app">
    <p v-for="(item,index) in list">
        {{index}}. {{item}}
    </p>
</div>
<script>
    new Vue({
        el: "#app",
        data: {
            list: ['HTML','CSS','JavaScript','Vue']
        }
    })
</script>
```

在浏览器中运行的效果如图 2.14 所示。

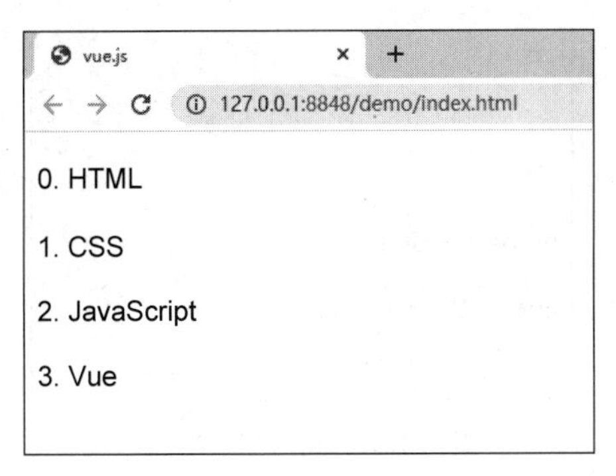

图 2.14　遍历数组效果

在 v-for 指令中，可以使用 of 替代 in 作为分隔符，因为它更接近 JavaScript 迭代器的语法，代码如下：

```
<div v-for="(item,index) of items">
    {{index}} - {{item}}
</div>
```

2. 遍历对象

v-for 指令不仅可以遍历数组，还可以用来遍历对象，代码如下：

```
<div v-for="(value,key) of object"></div>
```

在上面的示例代码中，使用 v-for 循环迭代出来的元素有两个参数，第一个参数为对象属性的值，第二个参数为对象的属性。

index.html 文件代码如下：

```
<div id="app">
    <p v-for="(value,key) in student">
        {{key}}: {{value}}
    </p>
</div>
<script>
    new Vue({
        el: "#app",
        data: {
            student: {
                id: 1,
                name: '韩梅梅',
                age: 18
            }
        }
    })
</script>
```

在浏览器中运行的效果如图 2.15 所示。

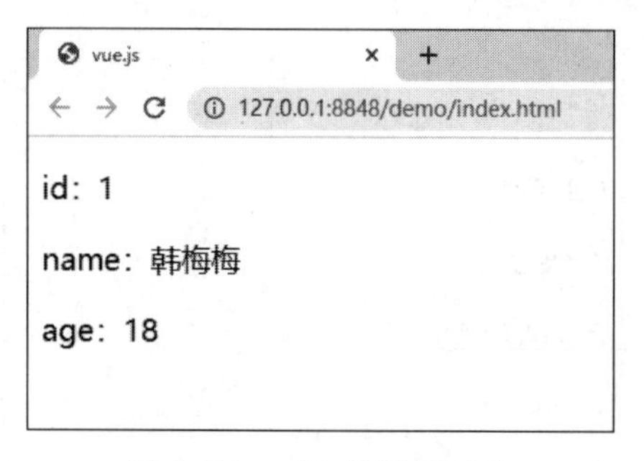

图 2.15　v-for 的遍历对象

使用 v-for 指令遍历对象时，迭代的元素使用第 3 个参数作为索引，在上面的代码中添加第 3 个参数，代码如下：

```
<div id = "app">
    <p v - for = "(value,key,index) in student">
        {{index}}. {{key}}: {{value}}
    </p>
</div>
```

在浏览器中运行的效果如图 2.16 所示。

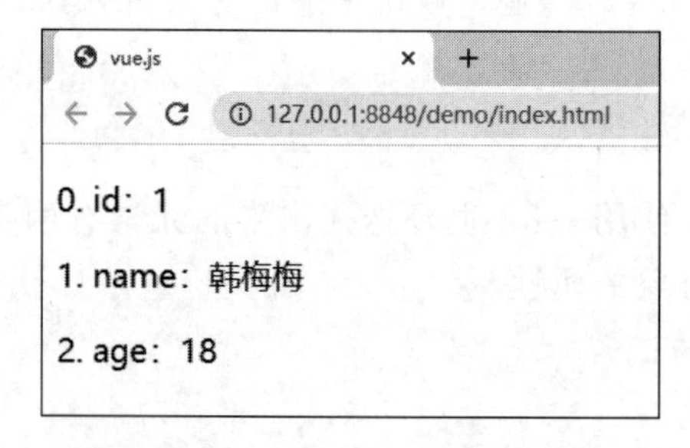

图 2.16　v-for 的第 3 个参数(索引)

3. 遍历整数

v-for 还可以直接遍历整数，代码如下：

```
<div id = "app">
    <p v - for = "i in 5">{{i}}</p>
</div>
```

在浏览器中运行的效果如图 2.17 所示。

2.3.2　维护状态

当 Vue 正在更新使用 v-for 渲染的元素列表时，它默认使用"就地更新"的策略。如果数据项的顺序被改变，Vue 将不会移动 DOM 元素来匹配数据项的顺序，而是就地更新每个元素，并且确保它们在每个索引位置被正确渲染。这类似于 Vue 1.x 的语句 track-by="$index"。

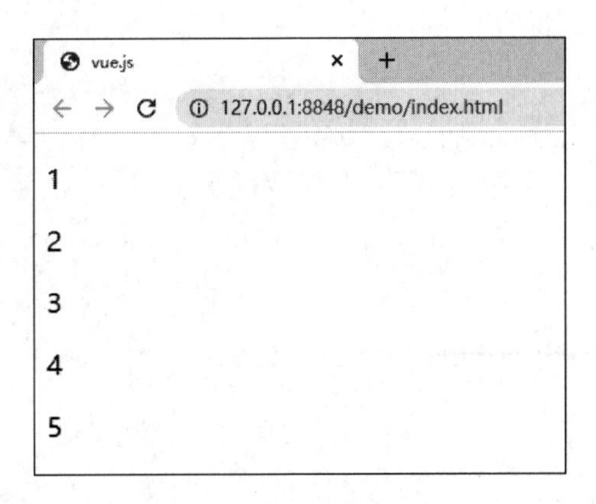

图 2.17　v-for 遍历整数

这个默认的模式是高效的，但是只适用于不依赖子组件状态或临时 DOM 状态（例如：表单输入值）的列表渲染输出。

为了给 Vue 一个提示，以便它能跟踪每个节点的身份，从而重用和重新排序现有元素，需要为每项提供一个唯一 key 属性，代码如下：

```
<p v-for="item in items" v-bind:key="item.id"></p>
```

建议尽可能在使用 v-for 时提供 key 属性，除非遍历输出的 DOM 内容非常简单，或者刻意依赖默认行为以获取性能上的提升。

注意　不要使用对象或数组之类的非基本类型值作为 v-for 的 key，而要使用字符串或数值类型的值。

2.3.3　数组更新检测

Vue 为了增强列表渲染功能，增加了一组观察数组的方法，并且可以显示一个数组的过滤或排序的副本。

1. 变更方法

Vue 将被侦听的数组的变更方法进行了包裹，所以它们也会触发视图更新。这些被包裹过的方法如下。

(1) push()：接收任意数量的参数并逐个追加到原数组末尾，返回新数组的长度。

(2) pop()：移除数组最后一项，返回被移除的元素。

(3) shift()：移除数组的第一项，返回被移除的元素。

(4) unshift()：在数组前追加新元素，返回新数组长度。

(5) splice()：删除指定索引的元素，并且可以在该索引处添加新元素。

(6) sort()：对数组进行排序，默认按字典升序排序，返回排序后的数组。

(7) reverse()：用于反转数组的顺序，返回反转后的数组。

这些方法类似于 JavaScript 中操作数组的方法。

index.html 文件代码如下：

```
<div id="app">
    <p v-for="item in list">{{item}}</p>
</div>
<script>
    var vm = new Vue({
            el: "#app",
            data: {
                    list: ['HTML','CSS','JavaScript','Vue']
            }
    })
    //使用 push()方法追加新元素
    vm.list.push('Jquery');
</script>
```

在浏览器中运行的效果如图 2.18 所示。

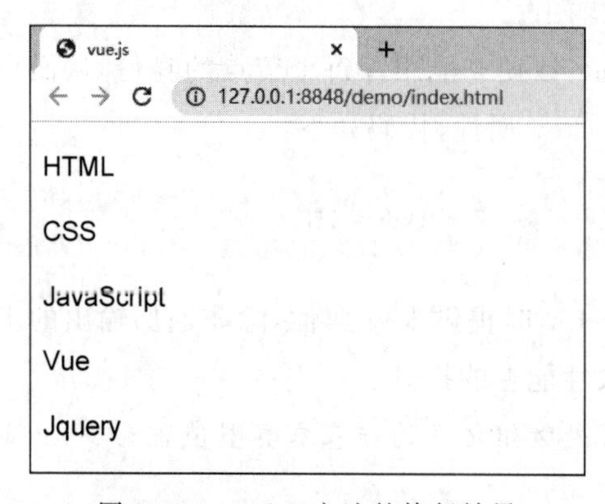

图 2.18　push()方法的执行效果

2. 替换数组

变更方法，顾名思义，会变更调用这些方法的原始数组。相比之下，也有非变更方法，例如 filter()、concat() 和 slice()。它们不会变更原始数组，而是返回一个新数组。当使用非变更方法时，可以用新数组替换旧数组。

非变更方法如下。

(1) concat()：创建当前数组的副本，然后拼接参数中的数组，返回拼接后的新数组。

(2) slice()：将数组的索引作为参数，可从已有的数组中返回选定的元素，返回新数组。

(3) map()：对数组的每一项运行给定函数，返回每次函数调用的结果所组成的数组。

(4) filter()：对数组的每一项运行给定函数，该函数会返回值为 true 的项所组成的数组。

非变更方法也和 JavaScript 中的方法类似。

index.html 文件代码如下：

```
<div id="app">
    <p v-for="item in items">{{item}}</p>
</div>
<script>
```

```
    var vm = new Vue({
        el: "#app",
        data: {
            numbers: [1,2,3,4,5,6]
        },
        computed:{
            items: function(){
                return this.numbers.filter(function(number){
                    return number < 4
                })
            }
        }
    })
</script>
```

在浏览器中运行的效果如图 2.19 所示。

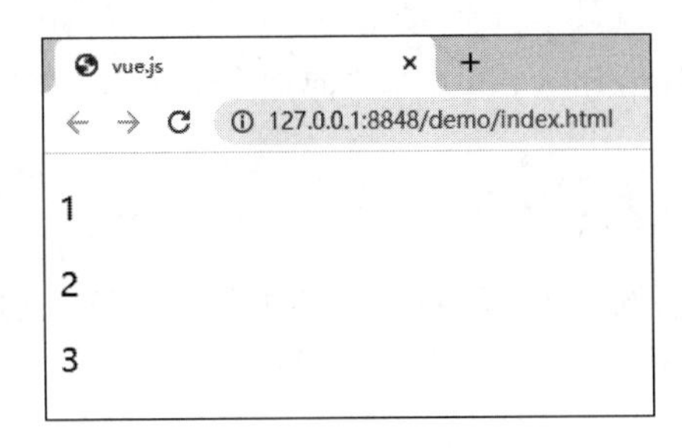

图 2.19　filter()方法的执行效果

在上面的代码示例中，要显示一个数组的过滤或排序副本，而不实际改变或重置原始数据(使用非变更方法)，可以使用 filter()方法。

2.3.4　对象变更检测注意事项

由于 JavaScript 的限制，Vue 无法检测对象属性的添加或移除。由于 Vue 会在初始化实例时对属性执行 getter/setter 转化，所以属性必须在 data 对象上存在才能让 Vue 将它转换为响应式的，这是由 Vue 的双向数据绑定决定的。

index.html 文件代码如下：

```
<div id="app">
    {{student.name}}, {{student.age}}
</div>
<script>
    var vm = new Vue({
        el: "#app",
        data: {
            student: {
                name: '韩梅梅'
            }
        }
```

```
    })
    //添加的 age 属性不是响应式的
    vm.student.age = 18
</script>
```

在上面的示例代码中添加的 age 属性不是响应式的，所以在页面中不会被渲染出来，在浏览器中运行的效果如图 2.20 所示。

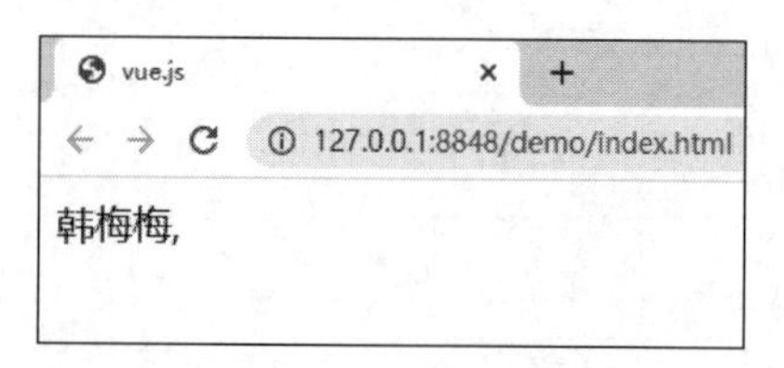

图 2.20　添加非响应式属性的效果

对于已经创建的实例，Vue 不允许动态添加根级别的响应式属性。但是，可以使用 Vue.set(object, propertyName, value) 方法向嵌套对象添加响应式属性，代码如下：

```
Vue.set(vm.someObject, 'b', 2)
```

我们可以使用上面的示例代码，为 Vue 实例中的对象添加响应式属性。

index.html 文件代码如下：

```
<div id="app">
    {{student.name}}, {{student.age}}
</div>
<script>
    var vm = new Vue({
        el: "#app",
        data: {
            student: {
                name: '韩梅梅'
            }
        }
    })
    //添加响应式属性
    Vue.set(vm.student, 'age',18)
</script>
```

在浏览器中运行的效果如图 2.21 所示。

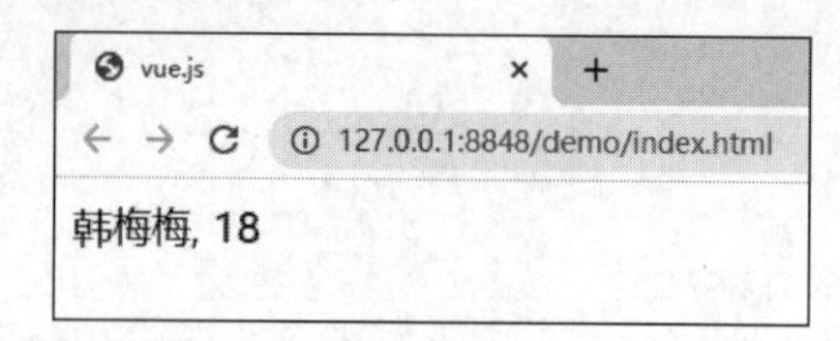

图 2.21　添加响应式属性效果

除了上面的方法，还可以使用 vm. $set 实例方法添加响应式属性，它只是全局 Vue. set 的别名，代码如下：

```
vm.$set(vm.student, 'age',18)
```

有时可能需要为已有对象赋值多个新属性，例如使用 Object. assign() 或 _. extend()。但是，这样添加到对象上的新属性不会触发更新。在这种情况下，应该用原对象与要混合进去的对象的属性一起创建一个新的对象，代码如下：

```
//代替 'Object.assign(this.someObject, { a: 1, b: 2 })'
this.someObject = Object.assign({}, this.someObject, { a: 1, b: 2 })
```

2.3.5 在<template>上使用 v-for

类似于 v-if，也可以利用带有 v-for 的 <template> 来循环渲染一段包含多个元素的内容。

index. html 文件代码如下：

```
<div id="app">
    <template v-for="item in list">
        <p>{{item}}</p>
    </template>
</div>
<script>
    var vm = new Vue({
        el: "#app",
        data: {
            list: ['HTML','CSS','JavaScript','Vue']
        }
    })
</script>
```

在浏览器中运行的效果如图 2.22 所示。

template 中可以放执行语句，最终编译后不会被渲染成元素。一般常和 v-for 指令及 v-if 指令结合使用，这样会使整个 HTML 结构没有多余的元素，从而使整个结构更加清晰。

2.3.6 v-for 与 v-if 一同使用

当它们处于同一节点，v-for 的优先级比 v-if 更高，这意味着 v-if 将分别重复运行于每个 v-for 循环中。当只想为部分项渲染节点时，这种优先级的机制十分有用。

index. html 文件代码如下：

```
<div id="app">
    <h3>年龄小于 20 岁的学生：</h3>
    <ul>
        <li v-for="stu in students" v-if="stu.age < 20">{{stu.name}}</li>
```

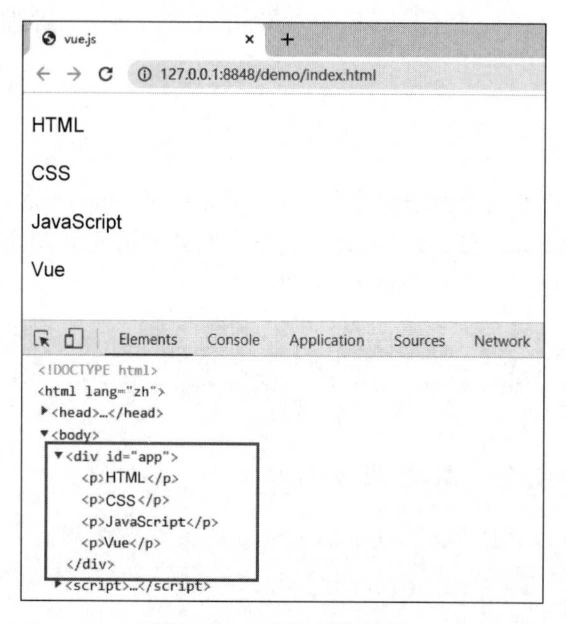

图 2.22　DOM 渲染结果

```
    </ul>
</div>
<script>
    var vm = new Vue({
        el: "#app",
        data: {
            students: [
                {name: '韩梅梅',age: 18},
                {name: '李雷',age: 19},
                {name: '大华',age: 20}
            ]
        }
    })
</script>
```

在浏览器中运行的效果如图 2.23 所示。

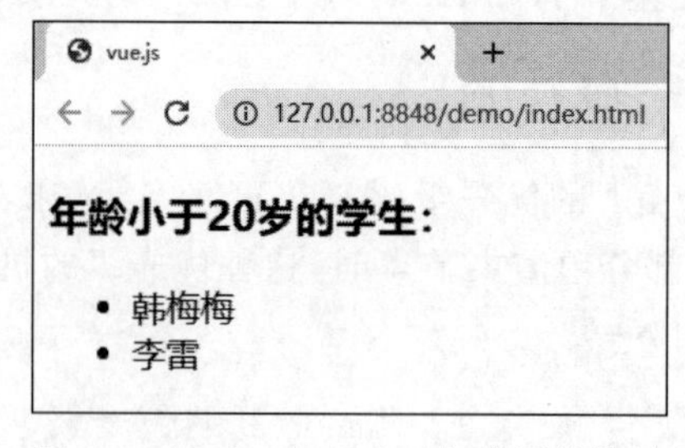

图 2.23　v-for 与 v-if 一起使用的效果

注意　不推荐在同一个元素上使用 v-for 和 v-if,必要情况下应该替换成计算属性 computed。

index.html 文件代码如下：

```
<div id = "app">
    <h3 >年龄大于 18 岁的学生：</h3 >
    <ul >
        <li v-for = "stu in student">{{stu.name}}</li >
    </ul >
</div >
<script >
    var vm = new Vue({
        el: "#app",
        data: {
            items: [
                {name: '韩梅梅',age: 18},
                {name: '李雷',age: 19},
                {name: '大华',age: 20}
            ]
        },
        computed: {
            student: function(){
                return this.items.filter(function(stu){
                    return stu.age > 18
                })
            }
        }
    })
</script >
```

在浏览器中运行的效果如图 2.24 所示。

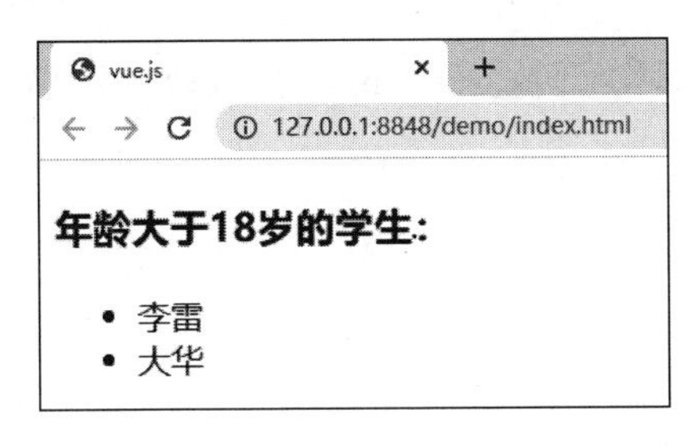

图 2.24　计算属性实现效果

2.4　事件处理

39min

2.4.1　监听事件

事件是指在浏览器中通过内置的处理器监视特定的条件或用户行为，例如鼠标单击浏览器窗口中的按钮。浏览器中内置了大量的事件处理器，当这些事件处理器被触发时，会执行一个绑定在该处理器上的函数，然后执行相应的内容。在 Vue 中可以使用 v-on 指令来完成事件函数的绑定。

index.html 文件代码如下：

```
<div id = "app">
    {{num}}
    <button v-on:click = "num ++">增加</button>
</div>
<script>
    var vm = new Vue({
        el: "#app",
        data: {
            num: 0
        }
    })
</script>
```

在浏览器中运行的效果如图 2.25 所示。

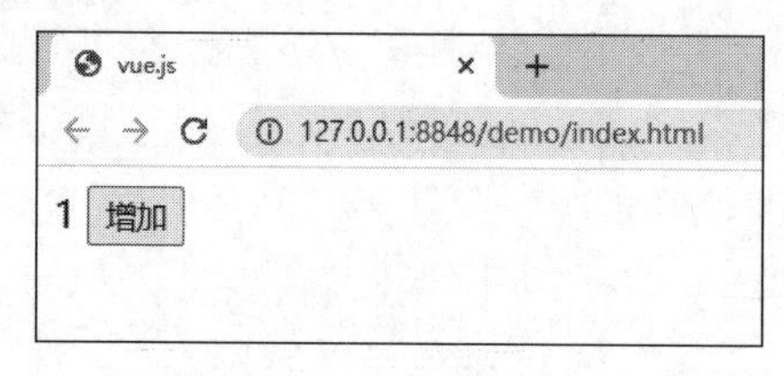

图 2.25　单击事件

在上面的示例代码中，单击“增加”按钮，num 属性的值会被加 1，并渲染到页面中。

2.4.2　事件处理方法

在 2.4.1 节的示例代码中，我们直接对 num 属性进行了操作，但是在实际的项目开发中，不能这样对属性进行直接操作。因为很多事件处理的逻辑比较复杂，应该把操作数据的代码写到具体的函数中。

index.html 文件代码如下：

```
<div id = "app">
    <button v-on:click = "reduce"> - </button>
    {{num}}
    <button v-on:click = "add"> + </button>
</div>
<script>
    var vm = new Vue({
        el: "#app",
        data: {
            num: 0
        },
        methods: {
```

```
            add: function(){
                this.num ++;
            },
            reduce: function(){
                this.num --;
            }
        }
    })
</script>
```

在浏览器中运行的效果如图 2.26 所示。

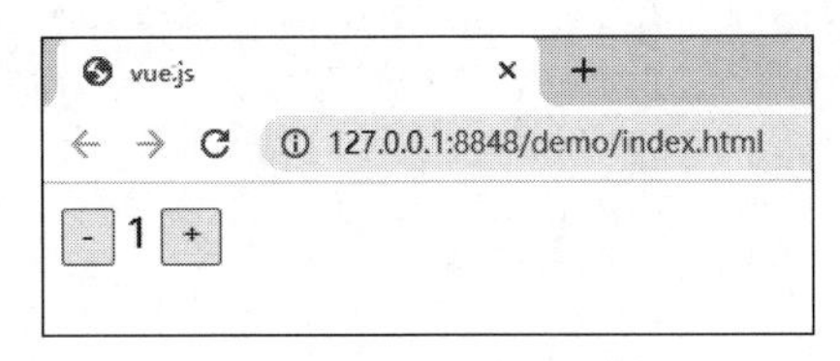

图 2.26　事件处理函数

注意　v-on 可以使用“@”代替，例如 <button @click="add">增加</button>。

在调用事件函数时，我们还可以为事件函数传入参数。

index.html 文件代码如下：

```
<div id="app">
    <button v-on:click="reduce(5)">-</button>
    {{num}}
    <button v-on:click="add(10)">+</button>
</div>
<script>
    var vm = new Vue({
        el: "#app",
        data: {
            num: 0
        },
        methods: {
            add: function(i){
                this.num += i;
            },
                this.num -= j;
            }
        }
    })
</script>
```

在浏览器中运行的效果如图 2.27 所示。

处理 <button> 标签，我们还可以在其他 DOM 元素上添加事件，如果在事件函数中要获取该事件所绑定的元素对象，可以使用事件函数的默认参数 $event 传参。

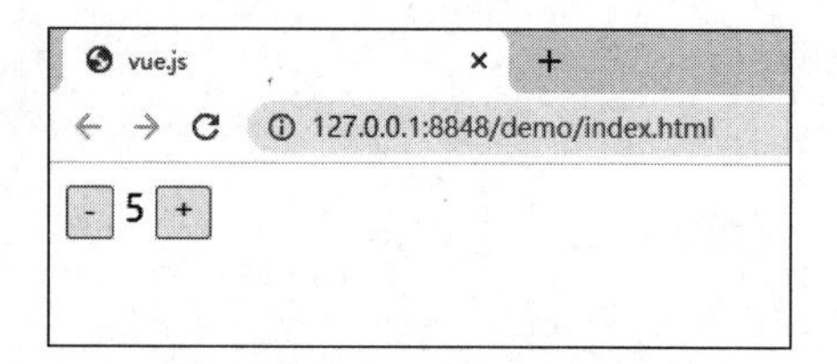

图 2.27　事件处理函数的传参

index.html 文件代码如下：

```
<div id="app">
    <input type="checkbox" @click="savePassword($event)"/>记住密码
</div>
<script>
    var vm = new Vue({
        el: "#app",
        methods: {
            savePassword(event){
                console.log(event)
            }
        }
    })
</script>
```

在浏览器中运行的效果如图 2.28 所示。

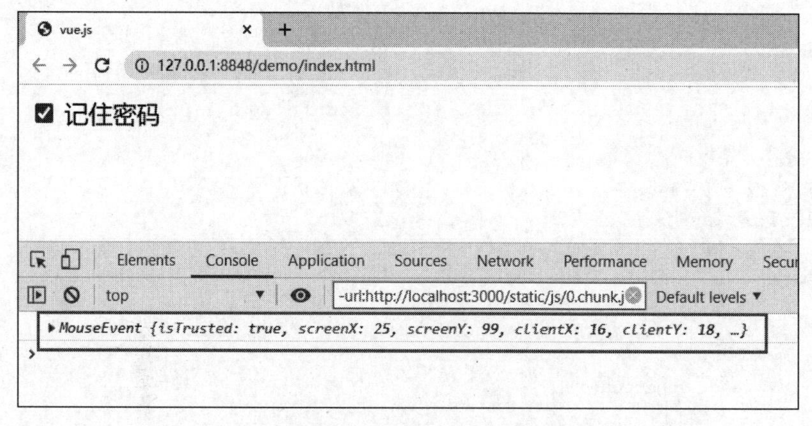

图 2.28　获取事件对象

event 对象代表事件的状态，例如触发事件所绑定的元素、当前键盘和鼠标按键的状态、鼠标当前所在的位置等。当一个事件被触发时，和当前这个事件相关的所有信息都会被保存到 event 对象中。

2.4.3　事件修饰符

在事件处理程序中调用 event.preventDefault() 或 event.stopPropagation() 是非常常见的需求。尽管我们可以在方法中轻松实现这点，但更好的方式是：方法只有纯粹的数据逻辑，而不是去处理 DOM 事件细节。

为了解决这个问题，Vue.js 为 v-on 提供了事件修饰符。使用修饰符可以节省很多代码和时间，这样便可以把更多的精力专注于处理程序的业务逻辑。

v-on 的修饰符是由点开头的指令后缀来表示的，在 Vue 中事件的修饰符主要有以下几个。

(1) .stop：等同于 JavaScript 中的 event.stopPropagation()，阻止事件冒泡。

(2) .self：只有触发当前修饰的元素时才会执行事件函数，不受事件冒泡影响。

(3) .capture：使用事件捕获模式，即内部元素触发的事件先在此处理，然后才交由内部元素进行处理。

(4) .once：只会触发一次。

(5) .prevent：等同于 JavaScript 中的 event.preventDefault()，阻止默认事件发生。

(6) .passive：执行默认行为。

1. stop 修饰符

stop 修饰符用来阻止事件冒泡，禁止事件继续向父级元素传播。例如，在评论区，当单击每条评论时，触发评论外层的 <div> 事件，当单击评论区内的用户头像时，可以查看用户的个人信息。如果按照事件冒泡机制，当单击评论区内的用户头像，即触发了头像上的时间，也会触发评论区 <div> 的事件。

index.html 文件代码如下：

```
<style type="text/css">
    /*设置评论区的样式 */
    .comment{
        width: 200px;
        height: 100px;
        border: 1px solid #000;
    }
    /*设置头像的样式 */
    .avatar{
        width: 50px;
        height: 50px;
        border: 1px solid #000;
        border-radius: 50%;
    }
</style>
<div id="app">
    <!-- 评论内容区 -->
    <div class="comment" @click="clickComment">
        <!-- 用户头像 -->
        <div class="avatar" @click="clickAvatar"></div>
    </div>
</div>
<script>
    var vm = new Vue({
        el: "#app",
        methods: {
            clickComment(){
```

```
                alert('评论区被单击了')
            },
            clickAvatar(){
                alert('头像被单击了')
            }
        }
    })
</script>
```

在浏览器中运行，单击头像会触发 clickAvatar()事件函数，效果如图 2.29 所示。根据事件冒泡机制，还会触发 clickComment()事件函数，效果如图 2.30 所示。

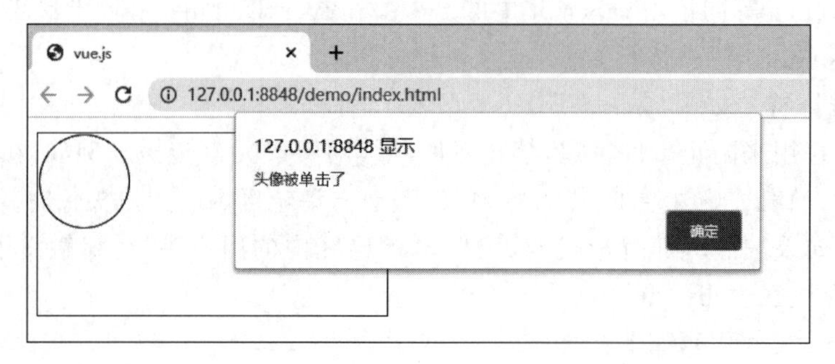

图 2.29　触发 clickAvatar 事件函数

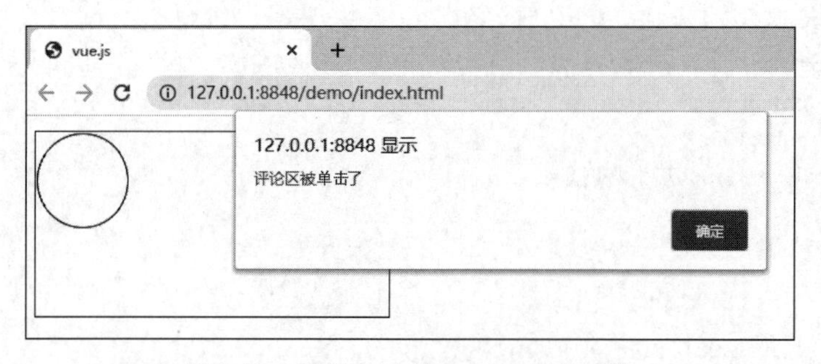

图 2.30　触发 clickComment 事件函数

上面示例代码中，单击头像会弹出两次提示框，在实际开发中，我们不需要在单击头像后触发外层的事件，此时可以使用 Vue 内置的修饰符 .stop 快速实现阻止事件冒泡的发生。

index.html 文件代码如下：

```
<div id="app">
    <!-- 评论内容区 -->
    <div class="comment" @click="clickComment">
        <!-- 用户头像 -->
        <div class="avatar" @click.stop="clickAvatar"></div>
    </div>
</div>
```

在浏览器中运行的效果如图 2.31 所示。

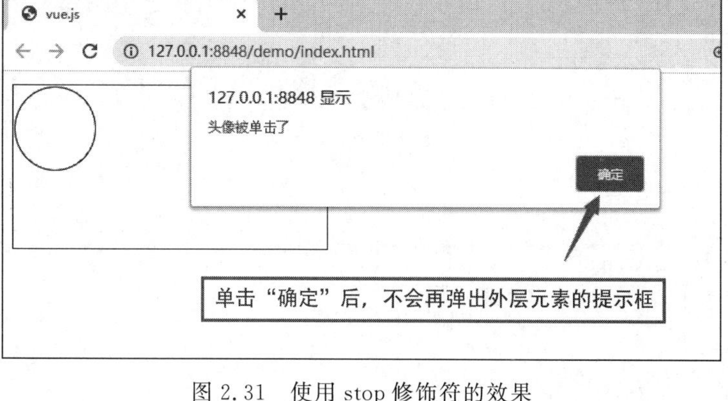

图 2.31　使用 stop 修饰符的效果

2. self 修饰符

self 修饰符可以跳过冒泡和捕获事件，只有当时触发了当前被修饰元素本身时，才执行绑定的事件函数。self 修饰符会监听事件是否直接作用在当前元素上。

index.html 文件代码如下：

```
<style type = "text/css">
    .outter{
        width: 150px;
        height: 150px;
        border: 1px solid #000;
    }
    .center{
        width: 100px;
        height: 100px;
        border: 1px solid #000;
    }
    .inner{
        width: 50px;
        height: 50px;
        border: 1px solid #000;
    }
</style>
<div id = "app">
    <div class = "outter" @click = "clickOutter">
        外层
        <div class = "center" @click.self = "clickCenter">
            中间层
            <div class = "inner" @click = "clickInner">
                内层
            </div>
        </div>
    </div>
</div>
<script>
    var vm = new Vue({
```

```
        el: "#app",
        methods: {
            clickOutter(){
                alert('外层被单击了')
            },
            clickCenter(){
                alert('中间层被单击了')
            },
            clickInner(){
                alert('内层被单击了')
            }
        }
    })
</script>
```

在浏览器中运行，当单击内层元素时，会触发内部绑定的 clickInner()事件函数，效果如图 2.32 所示。根据事件冒泡机制，会依次触发中间层和外层的事件函数，由于在中间层添加了 .self 修饰符，所以会跳过中间层，直接触发外层的事件函数，效果如图 2.33 所示。

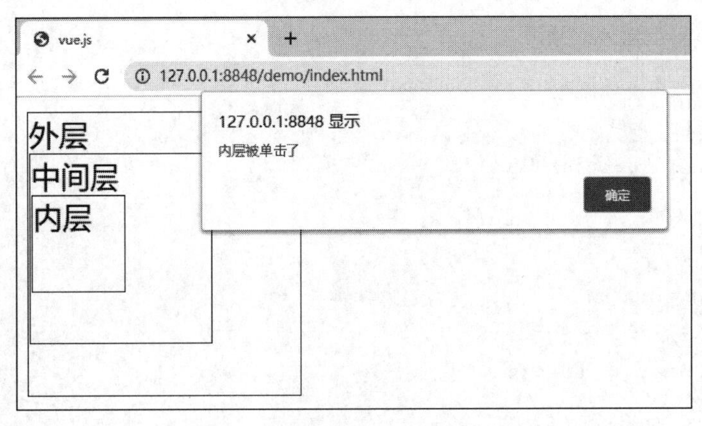

图 2.32　触发内层元素的事件函数

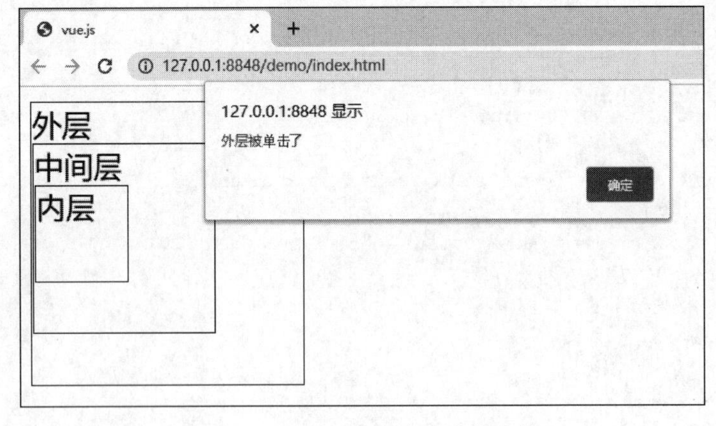

图 2.33　触发外层元素的事件函数

3. capture 修饰符

capture 修饰符可以为元素添加事件监听器时使用事件捕获模式，即内部元素触发的事件先在此处理，然后才交由内部元素进行处理。使用 capture 修饰符可以改变事件冒泡的执行顺序，先执行添加了该修饰符的元素。

在前面的示例代码中，我们可以为中间层添加 capture 修饰符。

index.html 文件代码如下：

```
<div id = "app">
    <div class = "outter" @click = "clickOutter">外层
        <div class = "center" @click.capture = "clickCenter">中间层
            <div class = "inner" @click = "clickInner">内层</div>
        </div>
    </div>
</div>
```

在浏览器中运行，单击内层元素，应该先执行内层元素的事件函数，但此处先执行了中间层的 clickCenter()函数，效果如图 2.34 所示。当执行完中间层的函数后，才去执行内层元素的 clickInner()函数，效果如图 2.35 所示。最后执行外层元素的事件函数，如图 2.36 所示。

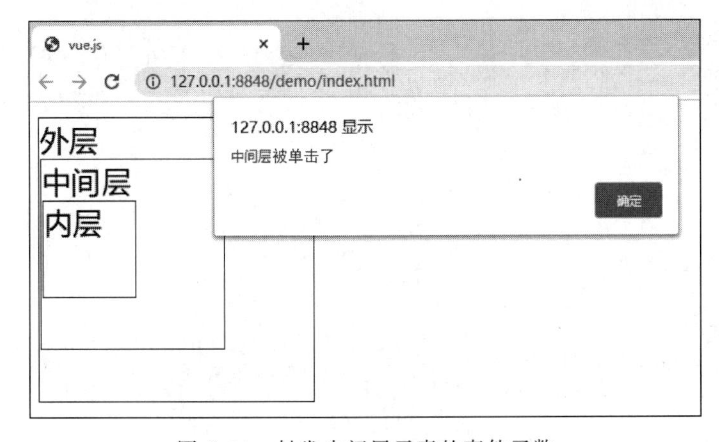

图 2.34　触发中间层元素的事件函数

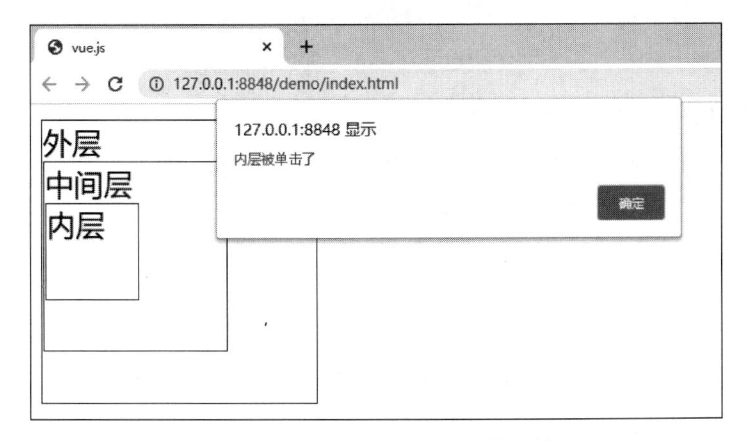

图 2.35　触发内层元素的事件函数

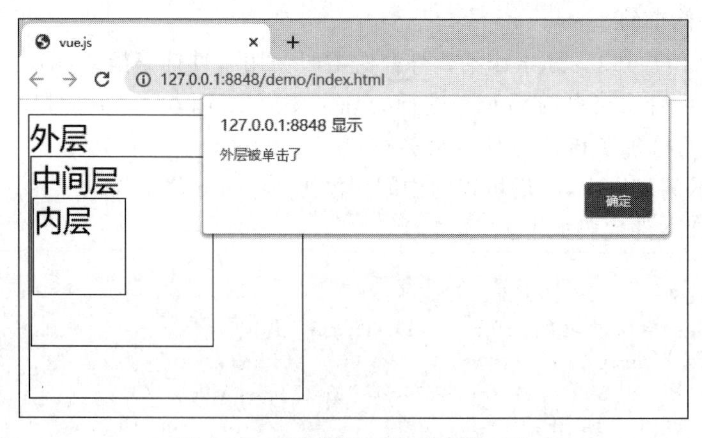

图 2.36　触发外层元素的事件函数

4. once 修饰符

有时我们需要对元素只执行一次操作，例如社交软件上的点赞操作，可以使用 once 修饰符来完成。

index.html 文件代码如下：

```
<div id="app">
    <button @click.once="clickStar">赞 {{star}}</button>
</div>
<script>
    var vm = new Vue({
        el: "#app",
        data: {
            star: 0
        },
        methods: {
            clickStar(){
                this.star ++;
            }
        }
    })
</script>
```

在浏览器中运行的效果如图 2.37 所示。

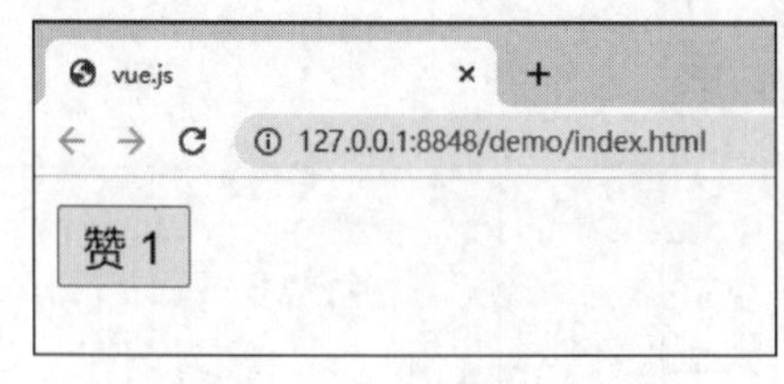

图 2.37　once 修饰符作用效果

5. prevent 修饰符

prevent 修饰符用于阻止浏览器的默认行为，例如 <a> 标签，使用 prevent 修饰符后，当单击超链接元素时，不会执行跳转动作。

index.html 文件代码如下：

```
<div id = "app">
    <a href = "http://cn.vuejs.org" @click.prevent = "clickLink"> Vue 官网</a>
</div>
<script>
    var vm = new Vue({
        el: "#app",
        data: {
            star: 0
        },
        methods: {
            clickLink(){
                alert('阻止了 a 标签跳转')
            }
        }
    })
</script>
```

在浏览器中运行的效果如图 2.38 所示。

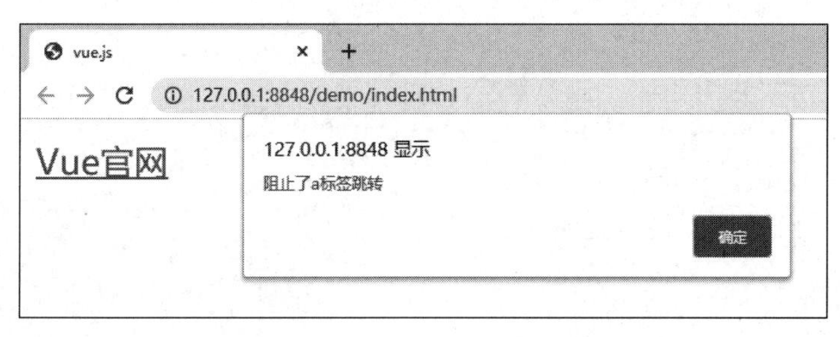

图 2.38 prevent 修饰符效果

6. passive 修饰符

passive 修饰符可以执行默认行为，元素本身的默认行为可以直接执行，为什么还要再添加一个 passive 修饰符来执行默认行为呢？这是因为浏览器只有当内核线程执行到事件监听器对应的 JavaScript 代码时，才能知道内部是否调用 preventDefault()函数来阻止事件的默认行为，所以浏览器本身是没有办法对这种场景进行优化的。这种场景下，用户的手势事件无法快速产生，这会导致页面无法快速执行滑动逻辑，从而让用户感觉到页面卡顿。

简单来说，当每次触发元素上的事件时，浏览器都会去查询一下是否有 preventDefault()函数阻止该次事件的默认行为。为元素添加 passive 修饰符，就是告诉浏览器不用再去检查了，表示该元素没有使用 preventDefault()函数阻止默认行为。

不要把 .passive 和 .prevent 一起使用，因为 .prevent 将会被忽略，同时浏览器可能会向你展示一个警告。需要记住，.passive 会告诉浏览器你不想阻止事件的默认行为。

注意 在使用修饰符时，修饰符的顺序很重要，相应的代码会以同样的顺序产生。因此，用 v-on:click.prevent.self 会阻止所有的单击，而 v-on:click.self.prevent 只会阻止对元素自身的单击。

2.4.4 按键修饰符

在 Vue 中支持 3 种键盘事件的监听。

（1）keydown：键盘按键按下时触发。

（2）keyup：键盘按键抬起时触发。

（3）keypress：键盘按键按下与抬起间隔期间触发。

在日常的开发中，我们经常需要用到键盘操作，例如，在搜索一个关键词之后，用户会习惯性地按下 Enter 键，以便于搜索结果。在传统的网页设计工作中，通常需要使用 JavaScript 中事件对象的 keyCode 属性来判断用户到底按下的是哪个键，然后进行后续操作。

为了方便开发，Vue 提供了在 v-on 监听事件时添加按键修饰符的方式，来监听用户的按键操作。

index.html 文件代码如下：

```
<div id="app">
    <input type="text" v-model="value" @keydown.a="handleInput" />
</div>
<script>
    var vm = new Vue({
        el: "#app",
        data: {
            value: ''
        },
        methods: {
            handleInput(){
                console.log('输入了 a')
            }
        }
    })
</script>
```

在浏览器中运行的效果如图 2.39 所示。

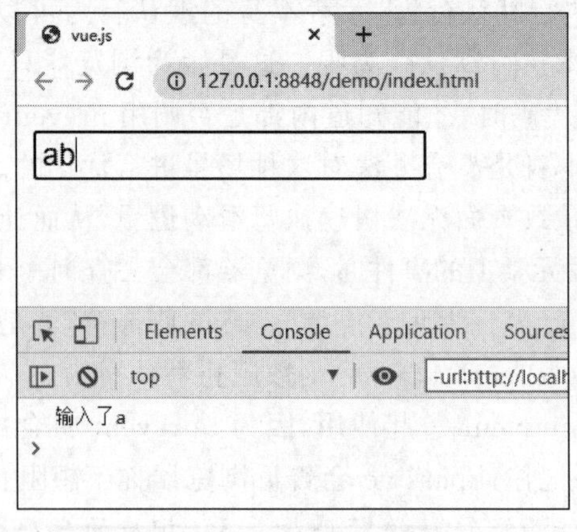

图 2.39 输入触发事件

在上面的示例代码中，为 <input> 输入框的 keydown 事件添加了 .a 修饰符，表示当前的输入框在输入字母 a 时会被监听，并触发绑定的事件函数。当在 <input> 输入框中输入字母 a 时会触发事件函数，但当输入字母 b 时不会触发事件函数。

为了方便使用按键修饰符，Vue 提供了绝大多数常用的按键码的别名，具体别名如下。

- .enter
- .tab
- .delete（捕获"删除"和"退格"键）
- .esc
- .space
- .up
- .down
- .left
- .right

对于上面的案例可以使用按键修饰符的别名，代码如下：

```
<input type="text" v-model="value" @keydown.a="handleInput" />
```

2.4.5 系统修饰键

可以使用下面的修饰符实现仅在按下相应按键时才触发鼠标或键盘事件的监听器。

- .ctrl
- .alt
- .shift
- .meta

index.html 文件代码如下：

```
<div id="app">
    <input type="text" v-model="value" @keydown.shift="handleInput" />
</div>
<script>
    var vm = new Vue({
        el: "#app",
        data: {
            value: ''
        },
        methods: {
            handleInput(){
                console.log('Shift 键被按下了')
            }
        }
    })
</script>
```

在浏览器中运行的效果如图 2.40 所示。

图 2.40　系统修饰键

2.4.6　为什么在 HTML 中监听事件

在前面各节的示例代码中，所有的事件都是通过在 HTML 标签中添加 v-on 属性实现的，这种事件监听的方式违背了关注点分离（Separation of Concern）这个长期以来的开发规范。之所以这样编写代码，是因为所有的 Vue.js 事件处理方法和表达式都严格绑定在当前视图的 ViewModel 上，它不会导致任何维护上的困难。实际上，使用 v-on 有以下几个特点。

（1）便于在 HTML 模板中快速定位 JavaScript 对应的事件函数。

（2）无须编写大量的 JavaScript 绑定事件的代码，可以更专注于编写业务逻辑。

（3）实现和 DOM 完全解耦，便于代码测试。

（4）当一个 ViewModel 被销毁时，所有的事件处理器都会被自动删除。

20min

2.5　表单输入绑定

2.5.1　双向绑定

MVVM 模式最重要的一个特性就是双向绑定，而 Vue 作为一个 MVVM 框架，也实现了数据的双向绑定。在 Vue 中使用内置的 v-model 指令完成数据在 View 与 Model 间的双向绑定。

v-model 会忽略所有表单元素的 value、checked、selected attribute 的初始值而将 Vue 实例的数据作为数据来源，因此应该通过 JavaScript 在组件的 data 选项中声明初始值。

2.5.2　v-model 基本用法

当 v-model 使用在不同的表单元素上时，保存值的类型也是不同的，常见的表单元素数据绑定操作如下。

1. 文本输入框

index.html 文件代码如下：

```
<div id="app">
    <input type="text" v-model="value"/>
    <p>value 的值：{{value}}</p>
</div>
<script>
    var vm = new Vue({
        el: "#app",
        data: {
            value: ''
        }
    })
</script>
```

在浏览器中运行的效果如图 2.41 所示。

图 2.41　绑定文本输入框

2. 文本域

index.html 文件代码如下：

```
<div id="app">
    <textarea rows="2" cols="15" v-model="message"></textarea>
    <p>文本域的值：{{message}}</p>
</div>
<script>
    var vm = new Vue({
        el: "#app",
        data: {
            message: ''
        }
    })
</script>
```

在浏览器中运行的效果如图 2.42 所示。

图 2.42　绑定文本域

3. 复选框

index.html 文件代码如下：

```
<div id="app">
    <input type="checkbox" v-model="hobby" value="篮球">篮球
    <input type="checkbox" v-model="hobby" value="足球">足球
    <input type="checkbox" v-model="hobby" value="排球">排球
    <p>爱好: {{hobby}}</p>
</div>
<script>
    var vm = new Vue({
        el: "#app",
        data: {
            hobby: []
        }
    })
</script>
```

在浏览器中运行的效果如图 2.43 所示。

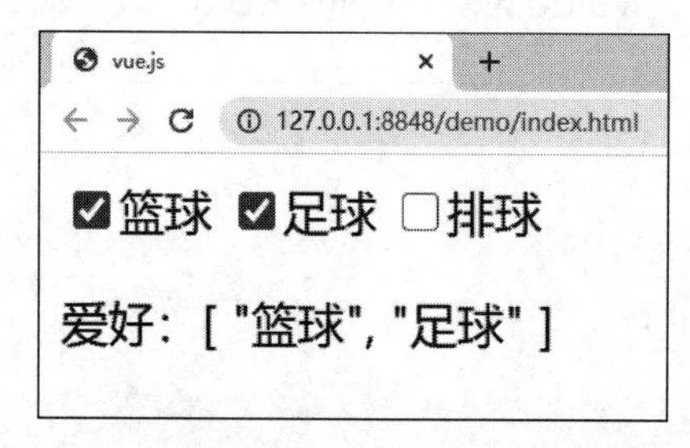

图 2.43　绑定复选框

4. 单选按钮

index.html 文件代码如下：

```
<div id="app">
    单选题:
    <input type="radio" v-model="selected" value="A">A
    <input type="radio" v-model="selected" value="B">B
    <input type="radio" v-model="selected" value="C">C
    <input type="radio" v-model="selected" value="D">D
    <p>选择: {{selected}}</p>
</div>
<script>
    var vm = new Vue({
        el: "#app",
        data: {
            selected: ''
        }
    })
</script>
```

在浏览器中运行的效果如图 2.44 所示。

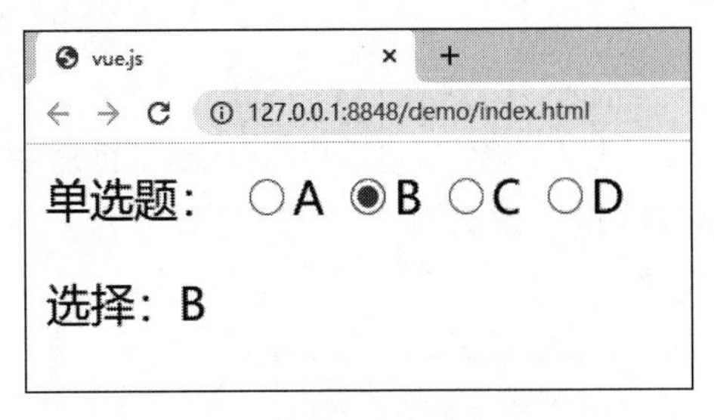

图 2.44　绑定单选按钮

5. 下拉选择框

index.html 文件代码如下：

```
<div id="app">
    年级：
    <select v-model="grade">
        <option value="一年级">一年级</option>
        <option value="二年级">二年级</option>
        <option value="三年级">三年级</option>
    </select>
    <p>选择：{{grade}}</p>
</div>
<script>
    var vm = new Vue({
        el: "#app",
        data: {
            grade: ''
        }
    })
</script></script>
```

在浏览器中运行的效果如图 2.45 所示。

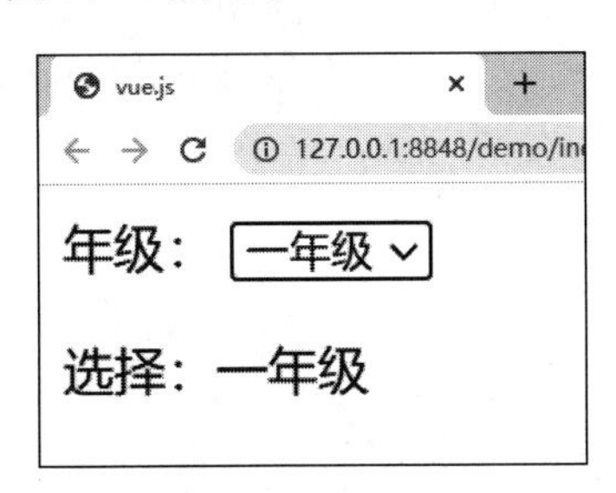

图 2.45　绑定下拉选择框

2.5.3　修饰符

对于 v-model 指令，有 3 种常用的修饰符。

- lazy

- number
- trim

1. lazy 修饰符

在输入框中，v-model 默认为同步数据，使用 lazy 修饰符后会转变为在 change 事件中同步，即当输入框失去焦点时数据才会更新。

index.html 文件代码如下：

```
<div id = "app">
    < input type = "text" v - model.lazy = "message" />
    < span >{{message}}</span >
</div >
< script >
    var vm = new Vue({
        el: "#app",
        data: {
            message: ''
        }
    })
</script >
```

在浏览器中运行，当输入内容时不会同步更新，如图 2.46 所示。当输入框失去焦点时，数据才会被更新，如图 2.47 所示。

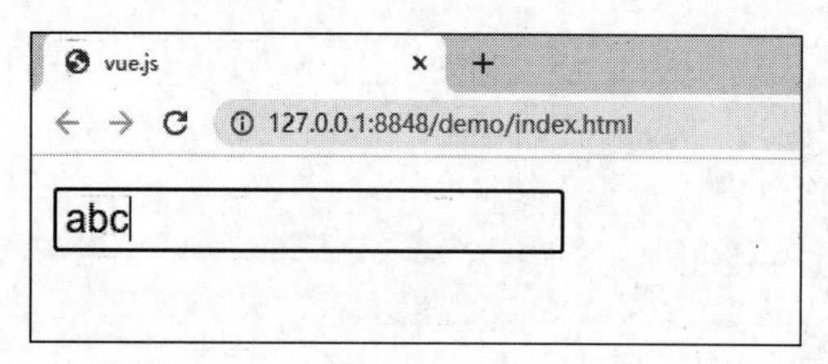

图 2.46 输入数据

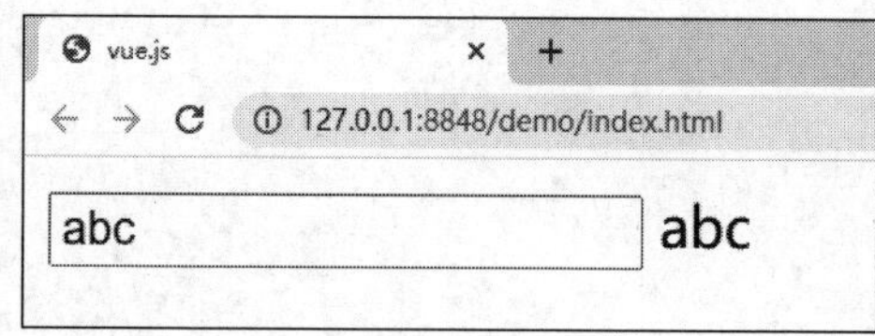

图 2.47 失去焦点时才更新数据

2. number 修饰符

输入框默认输入的值为 String 类型，使用 number 修饰符可以将输入的值转化为 Number 类型，该修饰符经常使用在数字输入框中。

index.html 文件代码如下：

```
< div id = "app">
    < input type = "text" v - model.number = "value" />
    < span >值的类型: {{typeof(value)}}</span >
</div >
< script >
    var vm = new Vue({
        el: "#app",
        data: {
            value: ''
```

```
        }
    })
</script>
```

在浏览器中运行，输入框的默认值为 String 类型，如图 2.48 所示。当输入数字时，值的类型会被转化为 Number 类型，如图 2.49 所示。

图 2.48　输入框的默认值类型

图 2.49　值的类型被转化为 Number 类型

3. trim 修饰符

如果要自动过滤用户输入内容的首尾空格，可以给 v-model 添加 trim 修饰符。

index.html 文件代码如下：

```
<div id="app">
    <input type="text" v-model.trim="value" />
    <span>值的长度：{{value.length}}</span>
</div>
<script>
    var vm = new Vue({
        el: "#app",
        data: {
            value: ''
        }
    })
</script>
```

在浏览器中运行的效果如图 2.50 所示。

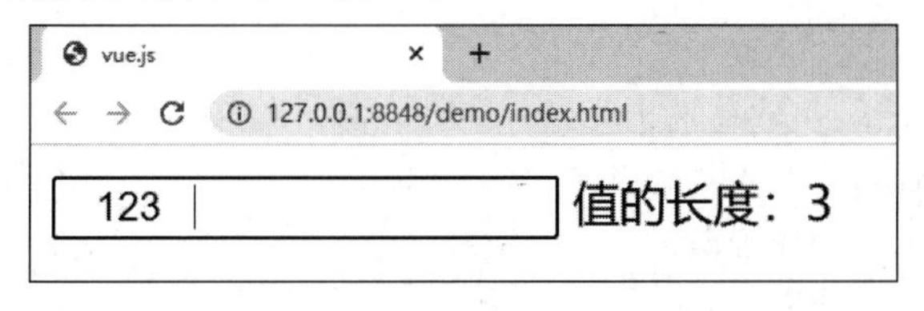

图 2.50　trim 修饰符效果

2.6 class 与 style 样式绑定

在 Vue 中，经常需要动态地设置元素的 CSS 样式，我们可以通过 v-bind 属性绑定，为标签的 class 属性和 style 属性动态赋值。

2.6.1 HTML 样式绑定

1. 数组语法

在 Vue 中，动态的样式类在 v-bind:class 中定义，静态的类名写在 class 样式中。

index.html 文件代码如下：

```
<style type="text/css">
    .box{
        width: 100px;
        height: 50px;
    }
    .style{
        border: 1px solid #000;
    }
</style>
<div id="app">
    <div v-bind:class="['box', 'style']">Hello World</div>
</div>
```

在浏览器中运行的效果如图 2.51 所示。

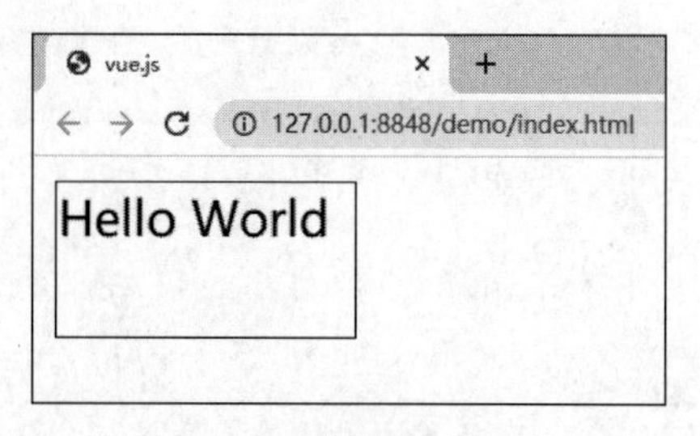

图 2.51 数组语法渲染

在上面的示例代码中，数组中样式的类名要使用单引号包裹，不然会出错。但是在某些应用场景下，需要使用变量表示类名，在数组中也可以使用这种方式绑定样式。

index.html 文件代码如下：

```
<style type="text/css">
    .box{
        width: 100px;
        height: 50px;
    }
    .style{
```

```
            border: 1px solid #000;
        }
    </style>
    <div id = "app">
        <div v-bind:class = "[ 'box', 'style' ]"> Hello World </div>
    </div>
    <script>
        var vm = new Vue({
            el: "#app",
            data: {
                Class1: 'box',
                Class2: 'style'
            }
        })
    </script>
```

在数组中也可以直接使用对象语法,示例代码如下:

```
<style type = "text/css">
    .box{
        width: 100px;
        height: 50px;
    }
</style>
<div id = "app">
    <div v-bind:class = "[ 'box',{border: '1px solid #000' }]"> Hello </div>
</div>
```

上面代码运行的效果和图 2.51 所示的效果是相同的。

2. 对象语法

在 Vue 中可以直接使用对象设置样式,对象的属性为样式的类名,其值为 Boolean 类型,当值为 true 时显示样式。对象的属性可以带引号,也可以不带引号。

index.html 文件代码如下:

```
<style type = "text/css">
    .box{
        width: 100px;
        height: 50px;
    }
    .style{
        border: 1px solid #000;
    }
</style>
<div id = "app">
    <div v-bind:class = "{box: b1, 'style': true}"> Hello World </div>
</div>
<script>
    var vm = new Vue({
```

```
        el: "#app",
        data: {
            b1: true
        }
    })
</script>
```

上面代码运行的效果和图 2.51 所示的效果相同。当对象中的属性过多时,如果将所有属性都写在元素上,则标签会显得代码冗余,此时我们可以使用计算属性来解决这个问题。

index.html 文件代码如下:

```
<style type="text/css">
    .box{
        width: 100px;
        height: 50px;
    }
    .style{
        border: 1px solid #000;
    }
</style>
<div id="app">
    <div v-bind:class="classObject">Hello World</div>
</div>
<script>
    var vm = new Vue({
        el: "#app",
        computed: {
            classObject: function(){
                return {
                    box: true,
                    style: true
                }
            }
        }
    })
</script>
```

上面代码运行的效果和图 2.51 所示的效果是相同的。

2.6.2 内联样式绑定

内联样式是将 CSS 样式编写到元素的 style 属性中,这种操作方法与设置 class 样式相同,也可以使用对象为元素进行设置。

index.html 文件代码如下:

```
<div id="app">
    <div v-bind:style="{color: styleColor, fontSize: fontSize + 'px'}">
        Hello World
```

```
    </div>
</div>
<script>
    var vm = new Vue({
        el: "#app",
        data: {
            styleColor: 'red',
            fontSize: 30
        }
    })
</script>
```

在浏览器中运行的效果如图 2.52 所示。

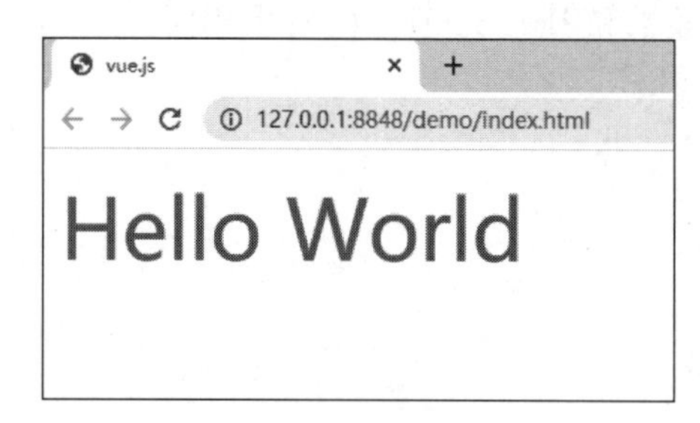

图 2.52　style 对象语法效果

同样，可以直接绑定一个样式对象属性，这样的代码看起来会更加简洁美观。

index.html 文件代码如下：

```
<div id="app">
    <div v-bind:style="styleObject">Hello World</div>
</div>
<script>
    var vm = new Vue({
        el: "#app",
        data: {
            styleObject: {
                styleColor: 'red',
                fontSize: 30
            }
        }
    })
</script>
```

还可以使用计算属性，返回样式对象，代码如下：

```
<div id="app">
    <div v-bind:style="styleObject">Hello World</div>
</div>
<script>
```

```
    var vm = new Vue({
        el: "#app",
        computed: {
            styleObject: function(){
                return {
                    color: 'red',
                    fontSize: '30px'
                }
            }
        }
    })
</script>
```

上面示例代码运行的效果和图 2.52 所示的效果相同。

第3章 Vue实例核心选项

3.1 数据选项

在实例化 Vue 对象时，需要为 Vue 的构造函数提供一系列配置信息，代码如下：

```
new Vue({
    //选项
})
```

当使用 new 操作符创建 Vue 实例时，可以为实例传入一个选项对象，选项对象中有很多类型的数据，具体内容如下。

3.1.1 data选项

data 选项支持 Object 和 Function 类型的数据，是 Vue 实例的数据对象。在 Vue 中使用递归将 data 的 property 转换为 getter/setter，从而让 data 的 property 能够响应数据变化。对象必须是纯粹的对象，代码如下：

```
var data = { a: 1 }

//直接创建一个实例
var vm = new Vue({
  data: data
})
vm.a // => 1
vm. $ data === data // => true
```

当定义组件时，data 必须声明为一个返回初始数据对象的函数。因为组件可能被用来创建多个实例，如果 data 仍然是一个纯粹的对象，则所有的实例将共享引用同一个数据对象。通过提供 data()函数，每次创建一个新实例后，可以通过调用 data()函数，返回初始数据的一个全新副本数据对象，代码如下：

```
//Vue.extend() 中 data 必须是函数
var Component = Vue.extend({
  data: function () {
```

```
    return{ a: 1 }
  }
})
```


如果在自定义组件中声明的 data 的 property 使用了箭头函数，那么 this 不会指向这个组件的实例，不过仍然可以将其实例作为函数的第一个参数访问，代码如下：

```
data:vm => ({ a: vm.myProp })
```

3.1.2 props 选项

props 选项的值可以是数组或对象，用于接收来自父组件的数据。props 可以是简单的数组，或者使用对象作为替代，对象允许配置高级选项，如类型检测、自定义验证和设置默认值。

props 选项的值如果为对象类型，则对象的语法具体使用如下。

（1）type：值的类型可以是 String、Number、Boolean、Array、Object、Date、Function、Symbol、任何自定义构造函数或上述内容组成的数组，可以通过该属性检查一个 prop 是否是指定的类型，否则会抛出警告。

（2）default：值为 any 类型，为该 prop 指定一个默认值。

（3）required：值为 Boolean 类型，定义该 prop 是否为必填项。

（4）validator：值为函数，自定义验证函数会将该 prop 的值作为唯一的参数代入。

props 选项代码如下：

```
//简单语法
Vue.component('props-demo-simple', {
  props: ['size', 'myMessage']
})

//对象语法，提供验证
Vue.component('props-demo-advanced', {
  props: {
    //检测类型
    height: Number,
    //检测类型 + 其他验证
    age: {
      type: Number,
      default: 0,
      required: true,
      validator: function (value) {
        return value >= 0
      }
    }
  }
})
```

3.1.3 computed 选项

computed 是计算属性，其值的类型是一个对象，对象中的属性值为函数。计算属性将被混入 Vue 实例中，所有 getter 和 setter 的 this 上下文自动绑定为 Vue 实例。如果计算属性的值为一个箭头函数，那么 this 不会指向这个组件的实例，不过仍然可以将其实例作为函数的第一个参数访问。

计算属性有以下特点。

（1）模板中放入太多的逻辑会让模板过重且难以维护，使用计算属性可以让模板更加简洁。

（2）计算属性是基于它们的响应式依赖进行缓存的。

（3）computed 比较适合对多个变量或者对象进行处理后返回一个结果值，也就是说多个变量中的某一个值发生了变化则我们监控的这个值也会发生变化。

index.html 文件代码如下：

```
<div id="app">
    <!--
        当多次调用 reverseString 的时候
        只要里面的 num 值不改变,它会把第一次计算的结果直接返回
        直到 data 中的 num 值改变,计算属性才会重新发生计算
    -->
    <div>{{reverseString}}</div>
    <div>{{reverseString}}</div>
      <!-- 调用 methods 中的方法的时候,它每次会重新调用 -->
    <div>{{reverseMessage()}}</div>
    <div>{{reverseMessage()}}</div>
  </div>
  <script type="text/JavaScript">
    /*
      计算属性与方法的区别:计算属性是基于依赖进行缓存的,而方法不缓存
    */
    var vm = new Vue({
      el: '#app',
      data: {
        msg: 'Nihao',
        num: 100
      },
      methods: {
        reverseMessage: function(){
          console.log('methods')
          return this.msg.split('').reverse().join('');
        }
      },
      //computed 属性的声明和 data 属性、methods 属性是平级关系
      computed: {
        //reverseString 这个是我们自己定义的名字
        reverseString: function(){
```

```
            console.log('computed')
            var total = 0;
            //当 data 中的 num 的值发生改变时,reverseString 函数会自动进行计算
            for(var i = 0;i <= this.num;i++){
              total  += i;
            }
            //在 reverseString 函数中必须有 return,否则在页面中无法获取计算后的值
            return total;
          }
        }
      });
    </script>
```

3.1.4 methods 选项

methods 用来定义 Vue 实例中绑定的函数,其值的类型为对象,对象的属性值必须为函数类型,methods 将被混入 Vue 实例中。可以直接通过 vm 实例访问这些函数,也可以在指令中直接使用。函数中的 this 自动绑定为 Vue 实例,代码如下:

```
var vm = new Vue({
  data:{ a: 1 },
  methods: {
    plus: function () {
      this.a++
    }
  }
})
vm.plus()
vm.a //2
```

15min

3.1.5 watch 选项

watch 是监听器,其值为一个对象,对象中的属性值可以为函数、对象和数组。当 data 中的一个属性需要被观察其内部值的改变时,可以通过 watch 来监听 data 属性的变化。

watch 监听器的基本语法如下。

(1) 使用 watch 来响应数据的变化。

(2) 一般用于异步或者开销较大的操作。

(3) watch 中的属性一定是 data 中已经存在的数据。

(4) 当需要监听一个对象的改变时,普通的 watch 方法无法监听到对象内部属性的改变,只有 data 中的数据才能够被监听到变化,此时需要 deep 属性对对象进行深度监听。

index.html 文件代码如下:

```
<div id = "app">
  <div>
    <span>名: </span>
```

```
      <span>
        <input type="text" v-model='firstName'>
      </span>
    </div>
    <div>
      <span>姓: </span>
      <span>
        <input type="text" v-model='lastName'>
      </span>
    </div>
    <div>{{fullName}}</div>
  </div>

  <script type="text/JavaScript">
    /* 侦听器 */
  var vm = new Vue({
      el: '#app',
      data: {
        firstName: 'Jim',
        lastName: 'Green',
        //fullName: 'Jim Green'
      },
      //watch 属性和 data 属性、methods 属性是平级关系
      watch: {
        //注意: 这里 firstName 对应 data 中的 firstName
        //当 firstName 值改变的时候会自动触发 watch
        firstName: function(val) {
          this.fullName = val + ' ' + this.lastName;
        },
        //注意: 这里 lastName 对应 data 中的 lastName
        lastName: function(val) {
          this.fullName = this.firstName + ' ' + val;
        }
      }
    });
  </script>
```

3.2 DOM 渲染选项

Vue 通过双向数据绑定,将数据实时渲染为浏览器能够解析的 DOM,在 Vue 实例中有一些关于操作 DOM 的选项,具体内容如下。

3.2.1 el 选项

el 可以是字符串也可以是一个节点,当使用字符串时 el 是 CSS 选择器,当使用节点时 el 是一个 HTMLElement 实例,代码如下:

```
<div id="app">
</div>
<script type="text/JavaScript">
    new Vue({
        el: '#app'
    })
</script>
```

3.2.2 template 选项

template 选项是一个字符串模板，属性值为字符串类型，作为 Vue 实例的标识使用。模板会替换挂载的元素，挂载元素的内容都将被忽略，除非模板的内容有分发插槽。

3.2.3 render 选项

render 是字符串模板的替代方案，可以使用 JavaScript 编程的方式实现。该渲染函数接收一个 createElement()方法作为第一个参数用来创建 VNode。如果组件是一个函数组件，则渲染函数还会接收一个额外的 context 参数，为没有实例的函数组件提供上下文信息。

25min

3.3 生命周期钩子

每个 Vue 实例都有完整的生命周期，即从开始创建、初始化数据、编译模板、DOM 挂载、数据更新并重新渲染、组件卸载等一系列过程，我们称为 Vue 实例的生命周期。在整个周期中，每个节点都会有一个钩子函数来处理该阶段的某些事物，这些钩子函数被称为生命周期函数。生命周期每个阶段具体的钩子函数内容如下。

3.3.1 create 初始化

1. beforeCreate

在 Vue 实例初始化之后，在数据观测和事件配置之前被调用，这时自定义组件的选项对象还没有被创建出来，el 和 data 选项并未初始化，所以在该钩子函数中无法访问 methods、data、computed 等选项上的方法和数据。

2. created

在 Vue 实例已经创建完成之后被调用，在该钩子函数中可以对 data 数据进行观测，对实例上的属性和方法进行运算，执行 watch 和 event 事件的回调，完成 data 数据的初始化等操作。但是在该阶段，由于还未到挂载阶段，所以 $el 属性仍然不能访问。

在 created 钩子函数中，我们常对一些实例进行预处理操作，例如发送 ajax 请求等。因为在该阶段可以调用 methods 中的方法，并对 data 中的数据进行修改。由于该阶段还未渲染 DOM，所以在该阶段中不能进行有关 DOM 操作的处理。

3.3.2 mount 组件挂载

1. beforeMount

在挂载之前被调用，Vue 中的 render()函数第一次被调用，在该阶段虚拟 DOM 已经完成了模板编译，把 data 中的数据和模板生成了 HTML，完成了 el 和 data 初始化工作。该阶段虽然已经完成了基本的初始化工作，但是还没有执行挂载操作。

2. mounted

在挂载完成后调用，也就是模板中的 HTML 已经被渲染到了浏览器中，一般会在该阶段执行 ajax 操作，或者执行 DOM 元素操作。

3.3.3 update 组件更新

1. beforeUpdate

在数据更新之前被调用，发生在虚拟 DOM 被重新渲染和打补丁之前，可以在该钩子中进一步更改状态，不会触发附加重复渲染过程。

2. updated

当 data 的数据发生改变时，虚拟 DOM 被重新渲染后会调用 updated()钩子函数。在调用时，组件 DOM 已经发生了更新，在这个阶段可以执行依赖 DOM 的操作。在该阶段应该避免出现操作数据的情况，因为可能会导致虚拟 DOM 被重新渲染，从而使更新进入无限循环的状态。该钩子函数在服务器端渲染期间不会被调用。

3.3.4 destroy 组件销毁

1. beforeDestroy

在 Vue 实例销毁之前调用，此时的实例还是完全可用状态。在该阶段可以使用 this 获取 Vue 实例，一般情况下会在该钩子函数中进行一些重置操作。例如，清除组件中的定时器，或清除监听 DOM 的事件等。

2. destroyed

在 Vue 实例被销毁之后调用，当钩子函数被调用后，所有的事件监听都会被移除，并且所有的组件都会被销毁。该钩子函数在服务器端渲染时不会被调用。

3.4 资源选项

3.4.1 directives 选项

7min

在 Vue 中除了内置的指令，例如 v-model 和 v-bind 等，Vue 还允许手动注册自定义指令。在 Vue 2 中，代码复用和抽象的主要形式是组件，然而在有些情况下，仍然需要对普通 DOM 元素进行底层操作，这时就需要用 directives 选项来自定义指令。

一个指令定义对象可以提供以下几个钩子函数。

(1) bind：只调用一次，指令第一次绑定到元素时调用。在这里可以进行一次性的初始化设置。

(2) inserted：被绑定元素插入父节点时调用（仅保证父节点存在，但不一定已被插入文档中）。

(3) update：所在组件的 VNode 更新时调用，但是可能发生在其子 VNode 更新之前。指令的值可能发生了改变，也可能没有。

(4) componentUpdated：指令所在组件的 VNode 及其子 VNode 全部更新后调用。

(5) unbind：只调用一次，指令与元素解绑时调用。

如果想要在页面渲染完成后自动让输入框获取焦点，可以使用 directives 选项创建一个自定义指令，代码如下：

```
directives: {
  focus: {
    //指令的定义
    inserted: function (el) {
      el.focus()
    }
  }
}
```

在组件中使用自定义指令，代码如下：

```
<input v-focus>
```

11min

3.4.2 filters 选项

Vue.js 允许自定义过滤器，可以被用于一些常见的文本格式化。过滤器可以用在两个地方：双花括号插值和 v-bind 表达式。过滤器应该被添加在 JavaScript 表达式的尾部，由"管道"符号指示，代码如下：

```
var vm = new Vue({
  el: '#app',
  data: {
    msg: ''
  },
  //filters 属性的声明和 data 属性、methods 属性是平级关系
  //定义 filters 中的过滤器为局部过滤器
  filters: {
    //upper 自定义的过滤器名字
    //upper 被定义为接收单个参数的过滤器函数，表达式 msg 的值将作为参数传入函数中
    upper: function(val) {
      //过滤器中一定要有返回值，这样外界使用过滤器的时候才能得到结果
      return val.charAt(0).toUpperCase() + val.slice(1);
    }
  }
});
```

过滤器可以串联使用，在此处是将 filter1 的值作为 filter2 的参数使用，代码如下：

```
{{data|filter1|filter2  }}
```

还可以为过滤器传递参数，例如，filterA 被定义为接收 3 个参数的过滤器函数，其中 message 的值作为第 1 个参数，普通字符串 'arg1' 作为第 2 个参数，表达式 'arg2' 的值作为第 3 个参数，代码如下：

```
<div id = "box">
  {{ message | filterA( 'arg1', 'arg2') }}
</div>
<script>
  //在过滤器中第 1 个参数对应的是管道符前面的数据 n,此时对应 message
  //第 2 个参数 a 对应实参 arg1 字符串
  //第 3 个参数 b 对应实参 arg2 字符串
  Vue.filter( 'filterA',function(n,a,b){
  if(n<10){
    return n+a;
  }else{
    return n+b;
  }
});
new Vue({
  el:"#box",
  data:{
    message: "哈哈哈"
  }
})
</script>
```

第4章 Vue工程化项目

随着多年的发展，前端越来越模块化、组件化、工程化，这是前端发展的大趋势。Webpack是目前用于构建前端工程化项目的主流工具之一，也正变得越来越重要。本章我们来详细讲解一下如何使用Webpack搭建Vue工程化项目。

4.1 使用Webpack构建Vue项目

4.1.1 什么是Webpack

Webpack是一个现代JavaScript应用程序的静态模块打包器（Static Module Bundler）。在Webpack处理应用程序时，它会在内部创建一个依赖图（Dependency Graph），用于映射到项目需要的每个模块，然后将所有这些依赖生成到一个或多个bundle。

Webpack可以做到按需加载。像Grunt、Gulp这类构建工具，打包的思路如下：

遍历源文件→匹配规则→打包

在这个过程中做不到按需加载，即对于打包的资源，到底页面用不用，在打包过程中是不关心的。

Webpack与其他构建工具本质上不同之处在于：Webpack从入口文件开始，经过模块依赖加载、分析和打包3个流程完成项目的构建。在加载、分析和打包的3个过程中，可以针对性地做一些解决方案，达到按需加载的目的，例如Code Split(拆分公共代码等)。

当然，Webpack还可以轻松地解决传统构建工具解决的问题：

- 模块化打包
- 语法糖转换
- 预处理器编译
- 项目优化
- 解决方案封装
- 流程对接

4.1.2 Webpack中配置Vue开发环境

在Webpack中配置Vue开发环境的过程如下。

1. 项目初始化

在硬盘上创建项目的根目录，例如d:\myapp，在myapp目录下启动命令行工具，执行命令如下：

```
npm init -y
```

上面的命令运行成功后，会在 myapp 目录下自动创建 package.json 文件。

2. 安装依赖

在 myapp 目录下的命令行工具中，依次执行命令如下：

```
#安装 Vue 依赖
npm i vue

#安装 Webpack 和 Webpack-cli 开发依赖
npm i Webpack Webpack-cli -D

#安装 Babel
npm i babel-loader @babel/core @babel/preset-env -D

#安装 Loader
npm i vue-loader vue-template-compiler -D

#安装 html-Webpack-plugin
npm i html-Webpack-plugin -D
```

3. 创建目录结构与文件

在 myapp 目录下依次新建 public 和 src 目录。在 public 目录下新建 index.html 文件，代码如下：

```
<!DOCTYPE html>
<html lang="zh-CN">
    <head>
        <meta charset="UTF-8" />
        <title>Webpack Vue Demo</title>
    </head>
    <body>
        <div id="app"></div>
    </body>
</html>
```

在 src 目录下分别新建 main.js 和 App.vue 文件。

main.js 文件代码如下：

```
import Vue from 'vue';
import App from './App.vue';

Vue.config.productionTip = false;

new Vue({
    render: h => h(App)
}).$mount('#app');
```

App. vue 文件代码如下：

```
<template>
    <div id="app">
        Hello Vue & Webpack
    </div>
</template>

<script>
    export default {};
</script>
```

4. 配置 Webpack. config. js

在 myapp 目录下新建 Webpack. config. js 配置文件，配置内容如下：

```
const HtmlWebpackPlugin = require('html-Webpack-plugin');
const VueLoaderPlugin = require('vue-loader/lib/plugin');

module.exports = {
    enter:'./src/main.js',
    resolve: {
        alias: {
            vue$: 'vue/dist/vue.esm.js'
        },
        extensions: ['*', '.js', '.vue', '.json']
    },
    module: {
        rules: [
            {
                test: /\.js$/,
                exclude: /node_modules/,
                use: {
                    loader: 'babel-loader'
                }
            },
            {
                test: /\.vue$/,
                loader: 'vue-loader'
            }
        ]
    },
    plugins: [
        new VueLoaderPlugin(),
        new HtmlWebpackPlugin({
            template: './public/index.html',
            filename: 'index.html'
        })
    ]
};
```

Vue 的配置文件中关于 Vue 语法的模板需要使用 vue-loader 来处理。完成上面配置后，执行 npx Webpack 命令，看一下 dist 输出的结果。

4.1.3 Webpack 配置本地服务器

Webpack-dev-server 是一个 Express 小型服务器，它通过 Express 的中间件 Webpack-dev-middleware 和 Webpack 进行交互。在开发过程中，如果项目本身就是个 Express 服务器，那么可以使用 Webpack-dev-middleware 和 Webpack-hot-middleware 两个中间件实现 HMR 功能。

Webpack-dev-server 具体操作如下。

1. 安装与启动

Webpack-dev-server 安装命令如下：

```
npm i Webpack - dev - server
```

安装成功后，执行以下命令启动本地服务器：

```
npx Webpack - dev - server
```

执行 Webpack-dev-server 命令之后，它会读取 Webpack 的配置文件（默认为 Webpack.config.js），然后将文件打包到内存中（所以看不到 dist 文件夹的生产，Webpack 会打包到硬盘上），这时打开 server 的默认网址：localhost:8080 就可以看到文件目录或者页面（默认为显示 index.html，如果没有此文件则显示目录）。

2. 自动刷新

在开发中，我们希望边写代码，边看到代码的执行情况，Webpack-dev-server 提供的自动刷新页面功能可以满足我们的需求。Webpack-dev-server 支持两种模式自动刷新页面。

（1）iframe 模式：页面被放到一个 iframe 内，当发生变化时，会重新加载。

（2）inline 模式：将 Webpack-dev-server 的重载代码添加到生成的 bundle 中。

两种模式都支持模块热替换（Hot Module Replacement）。模块热替换的好处是只替换更新的部分，而不是整个页面都被重新加载。

执行以下命令开启自动刷新：

```
Webpack - dev - server -- hot -- inline
```

42min

4.2 Vue CLI 脚手架工具

Vue CLI 是一个基于 Vue.js 进行快速开发的完整系统，致力于将 Vue 生态中的工具基础标准化。它确保各种构建工具能够基于智能的默认配置即可平稳衔接，这样就可以使开发者只专注于撰写应用，而不必浪费很多时间去研究项目搭建中的配置问题。

Vue CLI 有以下几个独立的部分。

1. CLI

CLI(@vue/cli)是一个全局安装的 NPM 包,提供了终端里的 vue 命令。它可以通过 vue create 命令快速搭建一个新项目,或者直接通过 vue serve 命令构建原型。

2. CLI 服务

CLI 服务(@vue/cli-service)是一个开发环境依赖,属于一个 NPM 包,局部安装在每个 @vue/cli 创建的项目中。CLI 服务构建于 Webpack 和 Webpack-dev-server 之上。

3. CLI 插件

CLI 插件是向 Vue 项目提供可选功能的 NPM 包,例如 Babel/TypeScript 转译、ESLint 集成、单元测试和 end-to-end 测试等。Vue CLI 插件的名字以 @vue/cli-plugin-(内建插件)或 vue-cli-plugin-(社区插件)开头,非常容易使用。

4.2.1 脚手架安装

Vue CLI 4.x 需要 Node.js v8.9 或更高版本(推荐 v10 以上版本)。更新依赖版本命令如下:

```
npm install -g @vue/cli
#或者
yarn global add @vue/cli
```

安装之后,就可以在命令行中访问 vue 命令。还可以通过运行一些简单的 vue 命令来验证它是否安装成功。

例如,可以执行命令来查看其版本,命令如下:

```
vue --version
```

如需升级全局的 Vue CLI 包,可以运行以下命令:

```
npm update -g @vue/cli
#或者
yarn global upgrade --latest @vue/cli
```

4.2.2 使用脚手架创建 Vue 项目

运行以下命令创建一个新项目:

```
#myapp 为项目名
vue create myapp
```

执行上面的命令会被提示选取一个 preset。可以选默认的包含了基本的 Babel+ESLint 设置的 preset,也可以选“手动选择特性”来选取需要的特性,如图 4.1 所示。

默认的设置非常适合快速创建一个新项目的原型,而手动设置则提供了更多的选项,它们是面向生产项目更加需要的。如果使用了手动选择特性,则可以在操作提示的最后选择将已选项保存为一个将来可复用的 preset。

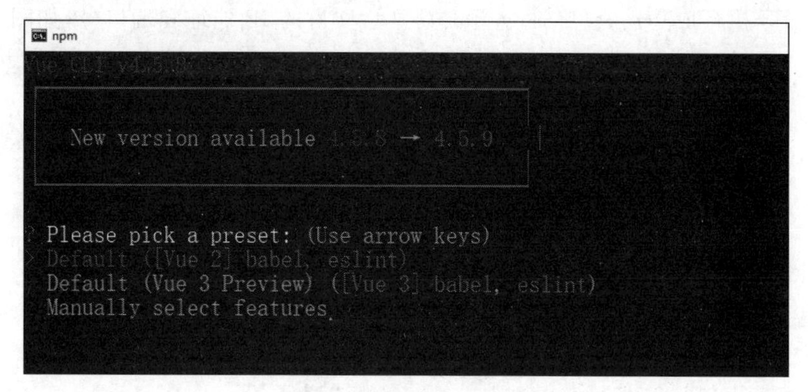

图 4.1　创建项目的提示

具体的选项操作可以参考图 4.2。

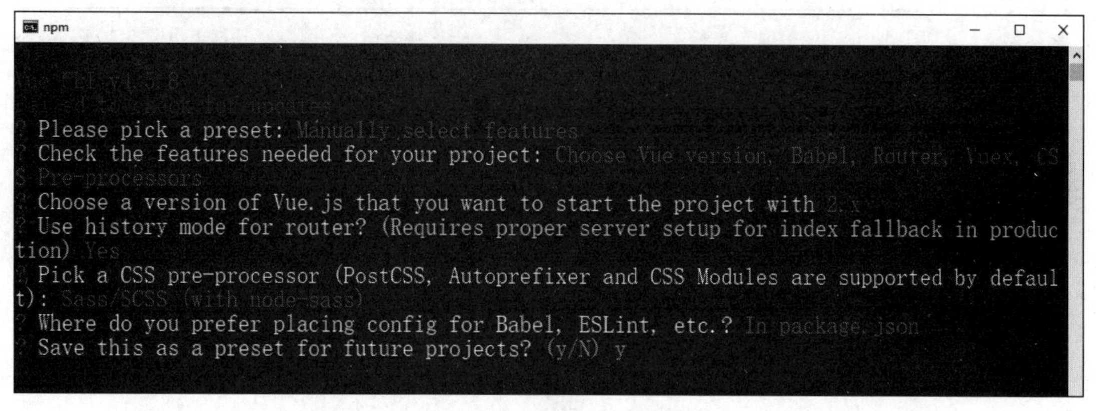

图 4.2　配置选项参考示例

上面的选项执行完成后，会提示如何启动项目，如图 4.3 所示。

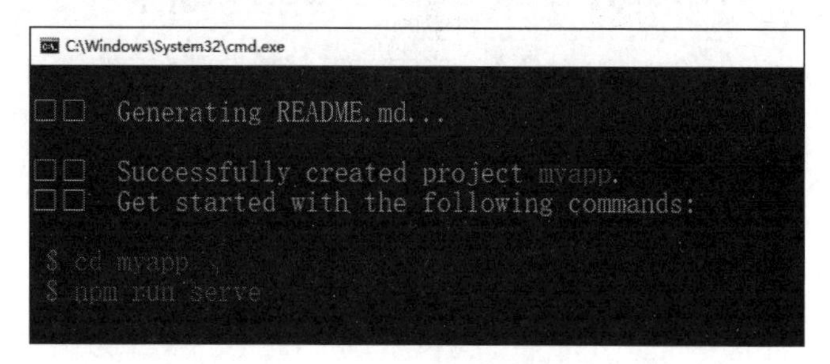

图 4.3　安装成功后的提示

根据命令行工具的提示，依次执行下面的命令，即可启动项目：

```
#进入项目目录
cd myapp
#启动项目
npm run serve
```

上面的命令执行成功后，会在命令行工具显示项目的 IP 地址和端口号，如图 4.4 所示。

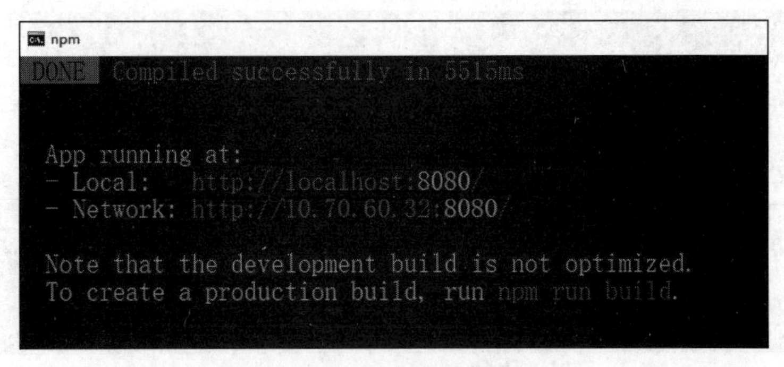

图 4.4　项目启动成功后的提示

在浏览器中访问 http://localhost:8080，打开项目的首页，效果如图 4.5 所示。

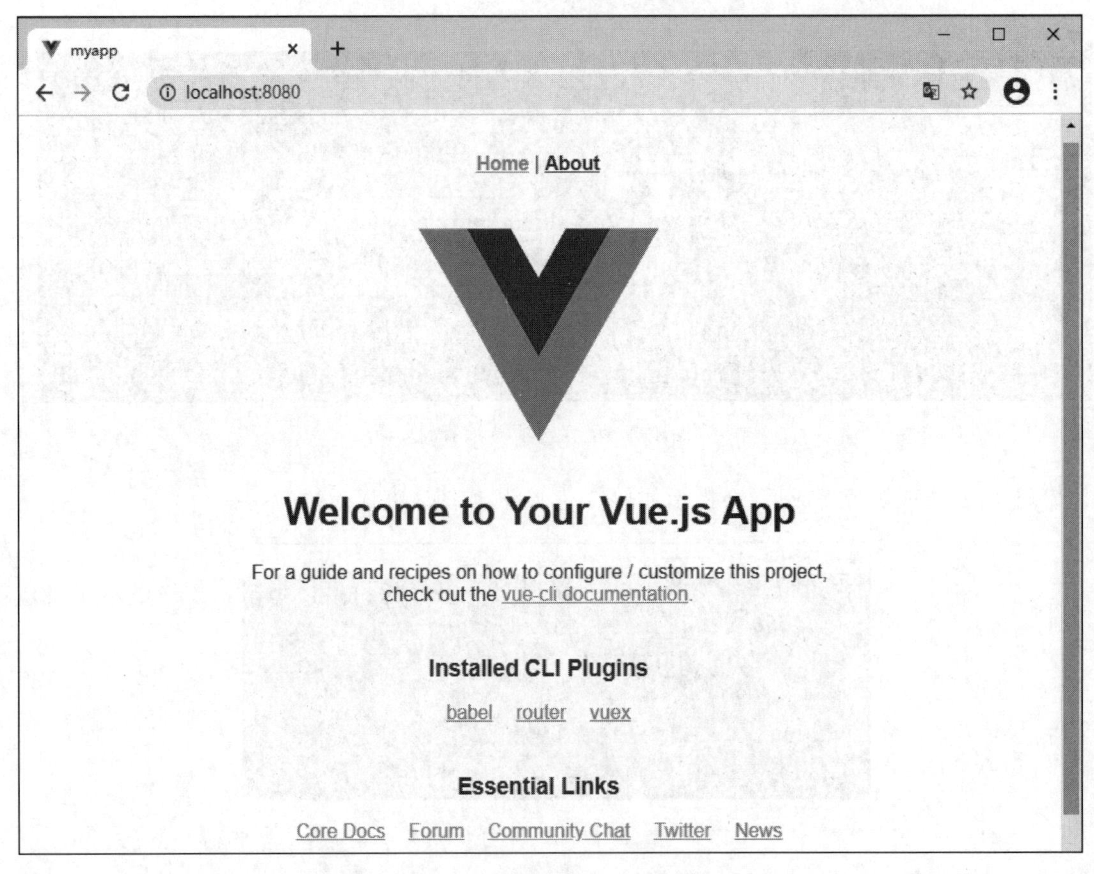

图 4.5　在浏览器中访问项目

4.2.3　项目结构与文件描述

把创建好的项目在 Visual Studio Code(以下简称 VSCode)开发工具中打开，项目的目录结构如图 4.6 所示。

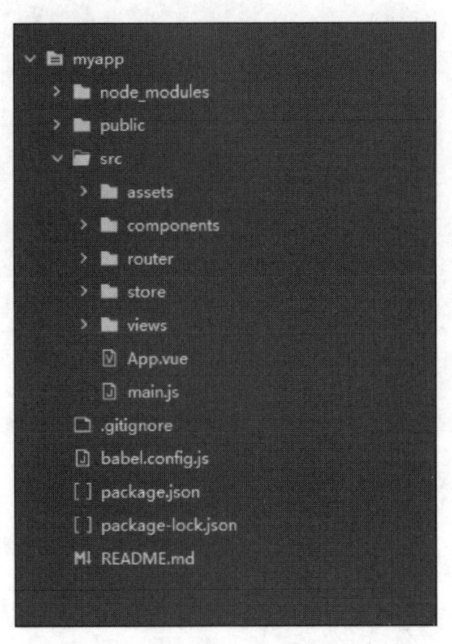

图 4.6　Vue.js 项目的目录结构

1. 目录结构解析

Vue CLI 4.0 版本脚手架工具创建的项目目录结构如表 4.1 所示。

表 4.1　项目目录结构

目录/文件	说　明
node_modules	NPM 加载的项目依赖模块
public	用于存放静态资源文件，启动本地服务器后，该目录为服务器访问的根目录。目录下存放 index.html 文件
src	这里是我们要开发的目录，基本上要做的事情都在这个目录里。里面包含了几个目录及文件： • assets：静态资源管理目录，如图片等 • components：公共组件管理目录 • router：路由配置文件目录 • store：状态管理文件目录 • views：视图组件管理目录 • App.vue：项目入口文件，也可以直接将组件写在这里，而不使用 components 目录 • main.js：项目的核心文件
babel.config.js	与 Babel 相关的配置文件
package.json	与 NPM 相关的配置文件
package-lock.json	与 NPM 相关的配置文件，用于锁定依赖包的版本号
README.md	Markdown 格式的项目说明文档

2. index.html 文件

public 目录下的 index.html 文件是项目的静态页面，当前创建的项目为单页面应用（SPA），所以整个项目中只有一个静态的 HTML 文件，index.html 文件代码如下：

```
<!DOCTYPE html>
<html lang="en">
  <head>
    <meta charset="UTF-8">
    <meta http-equiv="X-UA-Compatible" content="IE=edge">
    <meta name="viewport" content="width=device-width, initial-scale=1.0">
    <link rel="icon" href="<%= BASE_URL %>favicon.ico">
    <title><%= htmlWebpackPlugin.options.title %></title>
  </head>
  <body>
    <noscript>
      <strong>We're sorry but <%= htmlWebpackPlugin.options.title %> doesn't work
properly without JavaScript enabled. Please enable it to continue.</strong>
    </noscript>
    <div id="app"></div>
    <!-- built files will be auto injected -->
  </body>
</html>
```

3. main.js 文件

src 为项目的源码文件的管理目录，项目中编写的代码都要存放到这个目录下，在该目录下的 main.js 文件为整个项目的入口文件，关于项目的全局配置都要存放到该文件中，代码如下：

```
import Vue from 'vue'
import App from './App.vue'
import router from './router'
import store from './store'

Vue.config.productionTip = false

new Vue({
  router,
  store,
  render: h => h(App)
}).$mount('#app')
```

4. App.vue 文件

在 src 目录下还有一个很重要的文件，就是项目的根组件 App.vue 文件，该文件一般用于项目的一级路由的管理。因为在新建的项目中，会对 App.vue 编写初始化的页面内容，而且这些内容对后期的开发没有价值，所以在正式开发前，会对 App.vue 文件进行修改，代码如下：

```
<template>
  <div>
    <router-view/>
  </div>
</template>

<script>
  export default {}
</script>
```

第5章 深入了解 Vue 组件

5.1 什么是组件化开发

组件化开发是 Vue.js 框架的核心特性之一，也是目前前端技术框架中最常见的一种开发模式。在 Vue.js 中，组件就是一个可以复用的 Vue 实例，拥有独一无二的组件名称，可以扩展 HTML 元素，使用组件名称作为自定义的 HTML 标签。

在 Vue.js 项目中，每个组件都是一个 Vue 实例，所以组件内的属性选项都是相同的，例如 data、computed、watch、methods 及生命周期钩子等。仅有的例外是像 el 这样实例特有的选项。

在很多场景下，网页中的某些部分是可以复用的，例如头部导航、猜你喜欢、热点信息等。我们可以将网站中能够重复使用的部分设计成一个个组件，当需要的时候，直接引用这个组件即可。

Vue 组件化开发有别于前端传统的模块化开发。模块化是为了实现每个模块、方法的单一功能，一般通过代码逻辑进行划分，而组件化开发，更多的是实现前端 UI 的重复使用。

5.2 Vue 自定义组件

在使用 Vue CLI 工具创建的项目中，src 目录是用来存放项目源码的，在 src 目录下会自动创建两个子目录，一个是 src/views 目录，另一个是 src/components 目录。这两个子目录都是用来创建组件的，但是为了区分组件的功能，一般在 src/views 目录下创建的是视图组件，而在 src/components 目录下创建的是公共 UI 组件。

我们可以在 src/views 目录下创建一个 Home.vue 的视图组件，代码如下：

```
<template>
  <div>
    这是 Home 页面!
  </div>
</template>
```

然后将 Home.vue 组件引入项目的根组件 App.vue 中，代码如下：

```
<template>
  <div>
```

```
      <!-- 使用标签的方式使用自定义组件 -->
    <Home></Home>
  </div>
</template>

<script>
import Home from './views/Home.vue'
export default {
    components: {
            Home
    }
}
</script>
```

在 App.vue 根组件中使用 components 选项注册自定义组件，完成上面代码后启动项目，在浏览器中运行的效果如图 5.1 所示。

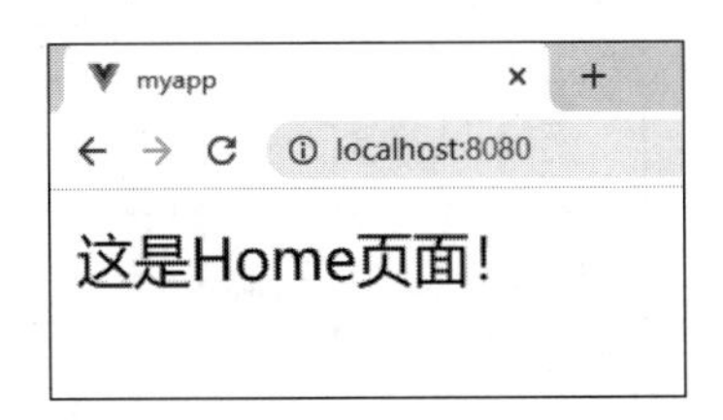

图 5.1　在浏览器中运行的效果

5.2.1　组件的封装

在项目的开发中，很多时候需要将某些 UI 部分封装成一个独立的组件，这部分 UI 可以是一个页面中的模块，例如商品列表，也可以是一个很小的部件，例如按钮。对于这种常用的 UI 元素可以创建一个组件并放到 components 目录下。

在 src/components 目录下新建一个 Button.vue 文件，代码如下：

```
<template>
    <button>按钮</button>
</template>
```

在上面的代码中，<template>标签内只能有一个子节点，如果在<button>标签的同级位置还有其他的标签元素，需要在<button>和同级标签的外部再添加一个父节点。

自定义 Button.vue 组件创建成功后，在 App.vue 根组件中引入，代码如下：

```
<template>
  <div>
      <!-- 用标签的方式使用自定义组件 -->
    <el-button />
  </div>
</template>
```

```
<script>
import Button from './components/Button.vue'
export default {
    components: {
            'el-button': Button
    }
}
</script>
```

在上面的代码中，components 选项内使用的 el-button 作为自定义组件的别名，在 <template>模板中用 <el-button> 作为 HTML 扩展标签。启动项目，在浏览器中运行的效果如图 5.2 所示。

图 5.2　自定义按钮组件

5.2.2　自定义组件上的属性

在 5.2.1 节示例代码中，使用 <el-button> 自定义 UI 组件就可以在浏览器中渲染出按钮元素，但是在很多时候，自定义组件内部不能直接定义按钮的样式与属性，应该由组件的使用者对按钮组件进行创建并赋予其相关的属性值。我们可以在自定义组件上以使用标签属性的方式，为自定义组件传入参数，并在组件内部渲染。

例如，可以在创建 Button.vue 组件时，为其设置 props 选项属性，代码如下：

```
<template>
    <button>{{text}}</button>
</template>

<script>
export default {
    props: {
        text: String
    }
}
</script>
```

在上面的示例代码中，使用 props 选项属性为 Button.vue 自定义组件定义了属性 text，在其他组件内引入该自定义组件时，就可以使用该属性作为标签属性，并为自定义组件传入具体的内容。

在 App. vue 中引入 Button. vue 自定义组件，代码如下：

```
<template>
  <div>
     <!-- 用标签的方式使用自定义组件，并通过 text 属性设置按钮文本内容 -->
    <el-button text="提交" />
  </div>
</template>

<script>
import Button from './components/Button.vue'
export default {
    components: {
            'el-button': Button
    }
}
</script>
```

在上面的示例代码中，通过为 <el-button> 标签设置 text 属性的方式设置按钮显示的文本内容，在浏览器中运行的效果如图 5.3 所示。

图 5.3　设置自定义按钮文本

5.2.3　自定义组件上的事件

自定义组件与预定义的 HTML 标签是有区别的，预定义的 HTML 标签能够被浏览器识别并渲染，所以会在浏览器中支持 HTML 标签上的原始属性和事件，但是自定义组件的标签在浏览器中渲染的是组件内部定义的 UI 样式，浏览器无法直接识别自定义组件标签上的属性和事件。

在使用自定义组件时，如果要为组件标签设置属性和事件，就要在自定义组件的内部提前声明。在组件上设置事件与设置属性还有所差别，设置属性只需要在组件内部使用 props 选项，而设置事件，需要考虑事件的执行时机，并在组件内部为原始的 HTML 标签定义该事件。

例如自定义按钮组件，并为组件设置单击事件，代码如下：

```
<template>
    <button @click="handleClick">{{text}}</button>
</template>
<script>
export default {
    props: {
        text: String
    },
    methods: {
        handleClick(){
```

```
                this.$emit('click')
            }
        }
    }
</script>
```

上面的示例代码为原生的 <button> 标签添加了单击事件，并在单击事件的触发函数中调用了 $emit()方法，触发该自定义组件定义的名为 click 的事件。

在 App.vue 中引入 Button.vue 组件，代码如下：

```
<template>
  <div>
      <!-- 用标签的方式使用自定义组件 -->
    <el-button text="提交" @click="submit"/>
  </div>
</template>
<script>
import Button from './components/Button.vue'
export default {
    components: {
        'el-button': Button
    },
    methods: {
        submit(){
            console.log('提交按钮被单击了')
        }
    }
}
</script>
```

在上面的示例代码中，为<el-button>标签定义了@click 事件，在自定义组件标签中没有默认的单击事件，但是在创建 Button.vue 组件时，为原生<button>按钮元素添加了单击事件，并使用 $emit('click') 方法触发了 click 事件，所以在<el-button>标签上定义@click 事件是可以被触发的。

在浏览器中运行的效果如图 5.4 所示。

图 5.4　自定义单击事件触发效果

5.3　组件属性校验

在自定义组件时，可以在自定义组件文件的内部使用 props 选项来为组件自定义一些功能。在使用自定义组件标签时，这些预先定义的 props 属性可以使用标签属性的书写方式在组件标签上声明，此时，标签上的属性值被传入自定义组件内的 props 选项中，在组件内就可以获取这个值了。

HTML 标签中的属性名对大小写不敏感，所以浏览器会把所有大写字符解释为小写字符。这意味着当使用 DOM 中的模板时，camelCase（驼峰命名法）的 prop 名需要使用其等价的 kebab-case（短横线分隔命名）命名，代码如下：

```
Vue.component('blog-post', {
  //在 JavaScript 中为 camelCase
  props: ['postTitle'],
  template: '<h3>{{ postTitle }}</h3>'
})
```

在 HTML 中属性名使用 kebab-case 格式命名，代码如下：

```
<blog-post post-title="hello!"></blog-post>
```

我们可以为组件的 prop 指定验证要求，如果有一个需求没有被满足，则 Vue 会在浏览器控制台中报出警告。在使用 prop 验证属性的类型时，值的类型应使用对象，而不能使用字符串，代码如下：

```
Vue.component('my-component', {
  props: {
    //基础的类型检查 ('null'和'undefined'会通过任何类型验证)
    propA: Number,
    //多个可能的类型
    propB: [String, Number],
    //必填的字符串
    propC: {
      type: String,
      required: true
    },
    //带有默认值的数字
    propD: {
      type: Number,
      default: 100
    },
    //带有默认值的对象
    propE: {
      type: Object,
      //对象或数组默认值必须从一个工厂函数获取
      default: function () {
```

```
        return { message: 'hello' }
      }
    },
    //自定义验证函数
    propF: {
      validator: function (value) {
        //这个值必须匹配下列字符串中的一个
        return ['success', 'warning', 'danger'].indexOf(value) !== -1
      }
    }
  }
})
```

33min

5.4 组件通信

组件化开发是Vue中的核心概念之一，通过设计具有各自状态的UI组件，然后由这些组件拼成更加复杂的UI页面，使代码更加简洁、容易维护。创建自定义组件在Vue开发中是非常常见的，在这种开发场景下必定会涉及组件之间的通信。在本节中将要学习的是如何实现组件之间的数据交互。

5.4.1 父组件向子组件通信

因为Vue项目采用单向数据流，所以只能从父组件将数据传递给子组件。具体传递的方式，可以使用5.2.2节讲到的props选项。在子组件的props选项中定义属性，在父组件中就可以向子组件传递值了。

子组件Son.vue文件代码如下：

```
<template>
    <div>
        子组件接收父组件传值：{{text}}
    </div>
</template>
<script>
export default {
    props: {
        text: String
    }
}
</script>
```

父组件Father.vue文件代码如下：

```
<template>
    <div>
        <h3>父组件向子组件传值</h3>
```

```
        <son v-bind:text="msg"></son>
    </div>
</template>
<script>
import Son from './Son.vue'
export default {
    components: {
        'son': Son
    },
    data(){
        return {
            msg: 'hello world!'
        }
    }
}
</script>
```

在浏览器中运行的效果如图 5.5 所示。

图 5.5 父组件向子组件传值

5.4.2 子组件向父组件通信

单向数据流决定了父组件可以影响子组件的数据，但是反之不行。子组件内数据发生更新后，在父组件中无法直接获取更新后的数据。要想实现子组件向父组件传递数据，可以在子组件数据发生变化后，触发一个事件方法，然后由这个事件方法告诉父组件数据更新了。在父组件中只需对这个事件进行监听，当捕获到这个事件运行后，再对父组件的数据进行同步更新。

子组件 Son.vue 文件代码如下：

```
<template>
    <div>
        子组件输入新值:
        <input type="text" v-model="value">
        <button @click="submit">提交</button>
    </div>
</template>
<script>
export default {
```

```
    data(){
        return {
            value: ''
        }
    },
    methods: {
        submit(){
            this.$emit('show',this.value);
        }
    }
}
</script>
```

父组件 Father.vue 文件代码如下：

```
<template>
    <div>
        <h3>父组件监听子组件数据更新：{{msg}}</h3>
        <son v-on:show="showMsg"></son>
    </div>
</template>
<script>
import Son from './Son.vue'
export default {
    components: {
        'son': Son
    },
    data(){
        return {
            msg: ''
        }
    },
    methods: {
        showMsg(msg){
            this.msg = msg
        }
    }
}
</script>
```

在上面的示例代码中，父组件中使用 v-on 事件监听器来监听子组件的事件，在子组件中使用 $emit() 触发当前实例上的事件。在浏览器中运行的效果如图 5.6 所示。

图 5.6　子组件向父组件传递数据

5.5 插　　槽

在构建页面时，我们常常会把具有公共特性的部分抽取出来，封装成一个独立的组件，但是在实际使用过程中又会产生一些其他问题，不能完全满足开发的需求。例如，当我们需要在自定义组件内添加一些新的元素时，原来组件的封装方式不能实现。这时，我们就需要使用插槽来分发内容。

5.5.1 什么是插槽

Vue 为了实现组件的内容分发，在组件的相关内容中提供了一套用于组件内容分发的 API，也就是插槽。这套 API 使用 <slot> 内置组件作为承载分发内容的出口，代码如下：

创建父组件 Demo.vue，代码如下：

```
<template>
    <div>
        <h3>在父组件中使用插槽</h3>
        <my-slot>
            <p>这是父组件中添加的元素</p>
        </my-slot>
    </div>
</template>
<script>
import Myslot from './Myslot.vue'
export default {
    components: {
        'my-slot': Myslot
    }
}
</script>
```

创建子组件 Myslot.vue，代码如下：

```
<template>
    <div>
        <p>这是子组件内容</p>
        <slot></slot>
    </div>
</template>
```

在浏览器中运行的效果如图 5.7 所示。

5.5.2 具名插槽

在 Vue 2.6 中，具名插槽和作用域插槽引入了一个新的统一语法的 v-slot 指令。它取代了 slot 和 slot-scope 特性。

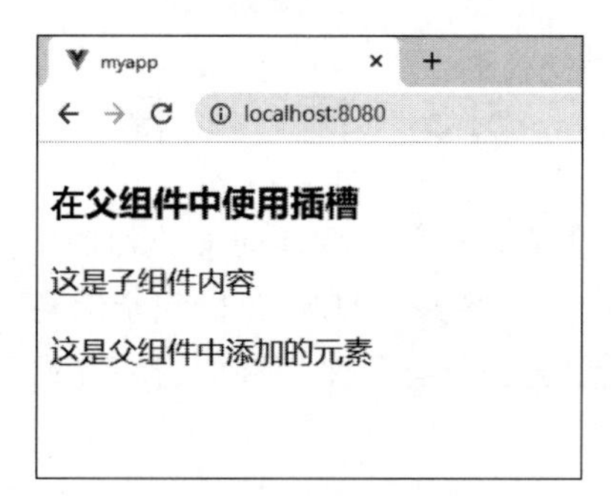

图 5.7　插槽效果

在实际的开发过程中，组件中的插槽不止一个，有时需要多个插槽，代码如下：

```
<div>
    <div class = "header">
        这是页面头部
    </div>
    <div>
        这是页面的主体内容
    </div>
    <div>
        这是页面的底部
    </div>
</div>
```

针对上面的示例，<slot>元素有一个 name 属性，这个属性可以用来定义额外的插槽，代码如下：

```
<div>
    <div class = "header">
        <slot name = "header"></slot>
    </div>
    <div>
        <slot></slot>
    </div>
    <div>
        <slot name = "footer"></slot>
    </div>
</div>
```

当<slot>元素上没有定义 name 属性时，<slot>出口会带有隐含的 default。在向具名插槽提供内容时，可以在一个<template>元素上使用 v-slot 指令，并以 v-slot 的参数形式提供其名称，代码如下：

```
<base - layout>
    <template v - slot:header>
        这是页面头部
    </template>
```

```
    <p>这是页面主体部分</p>
    <template v-slot:footer>
        这是页面底部
    </template>
</base-layout>
```

在上面的代码中，所有的内容都被传入对应的插槽，没有使用带有 v-slot 的<template>中的内容会被视为默认插槽的内容。

v-slot 指令与 v-on、v-bind 类似，也有自己的缩写形式，把 v-slot 替换为字符 # 即可，代码如下：

```
<base-layout>
    <template #header>
        这是页面头部
    </template>
    <p>这是页面主体部分</p>
    <template #footer>
        这是页面底部
    </template>
</base-layout>
```

5.5.3 作用域插槽

在使用插槽时，经常会有这样的应用场景：在父组件中定义的数据需要在子组件中也能访问。例如，在父组件中获取的商品列表数据，当父组件中引入商品卡片的组件时，需要在商品卡片组件内部使用插槽，并把商品数据传给子组件进行渲染。

例如，创建商品卡片组件 card.vue 文件，代码如下：

```
<template>
    <div>
        <slot>{{ goods.title }}</slot>
    </div>
</template>
```

在商品卡片组件中想要显示商品的标题，但是商品数据是在父组件中获取的，所以只有在父组件中使用<card>组件标签才可以访问 goods 对象。为了让 goods 在父级的插槽内容可用，可以将 goods 作为<slot>元素的一个属性绑定上去。

card.vue 文件代码如下：

```
<template>
    <div>
        <slot v-bind:goods="goods">
            {{ goods.title }}
        </slot>
    </div>
</template>
```

绑定在 <slot> 元素上的属性被称为插槽 prop。现在，在父级作用域中，可以给 v-slot 赋一个值，来定义提供的插槽 prop 的名字，代码如下：


```
<div>
    <card>
      <template v-slot:default="slotProps">
            {{ slotProps.goods.title }}
      </template>
    </card>
</div>
```

核心技术篇

第6章 Vue Router 路由

6.1 路由基础

33min

6.1.1 什么是路由

用 Vue. js 创建的项目是单页面应用，如果想要在项目中模拟出类似于页面跳转的效果，就要使用路由。其实，我们不能只从字面的意思来理解路由，从字面上来看，很容易把路由联想成“路由器”。路由器是连接两个或多个网络的硬件设备，而此处我们所说的路由，是指在一个应用程序中连接多个页面(组件)的一种配置。在一个全栈项目中，路由分为前端路由和后端路由。

1. 后端路由

先来看一下后端路由，例如项目的服务器网址是 http://192.168.1.10:8080，在这个站点中提供了 3 个界面，分别是：

页面 1，网址 http://192.168.1.10:8080/index.html

页面 2，网址 http://192.168.1.10:8080/about.html

页面 3，网址 http://192.168.1.10:8080/news.html

当在浏览器中输入 http://192.168.1.10:8080/index.html 时，Web 服务器接收到这个请求，然后把“/index.html”解析出来，再找到 index.html 文件并响应给浏览器，这就是服务器端的路由分发。

2. 前端路由

虽然前端路由和后端路由在实现技术上有些差别，但是实现的原理是一样的。在 HTML5 的 history API 发布之前，前端路由功能是通过哈希散列计算的，因为哈希算法可以兼容低版本的浏览器，例如：

http://192.168.1.10:8080/#/index.html

http://192.168.1.10:8080/#/about.html

http://192.168.1.10:8080/#/news.html

由于 Web 服务不会解析#后面的内容，而 JavaScript 可以获取#后面的内容，那么就可以使用 window.location.hash 来读取，通过这种方法来匹配到不同的功能上。使用哈希的方式还有一个很大的优点，当哈希的值改变后，不会导致浏览器的刷新。

6.1.2 在 Vue 中使用路由

用 Vue. js+Vue Router 创建单页应用非常简单。要在 Vue. js 应用程序中使用路由，需要先安装 vue-router，在当前项目下启动命令行工具，命令如下：

```
npm install vue - router
```


如果在一个模块化工程中使用它，必须通过 Vue.use()明确安装路由功能：

```
import Vue from 'vue'
import VueRouter from 'vue - router'

Vue.use(VueRouter)
```

如果使用脚手架工具创建项目，则路由的配置在 /src/router/index.js 文件中。

注意　如果使用全局的 script 标签，就无须上面的操作了。

在脚手架工具创建的项目中使用路由，需要在 /src/router/index.js 路由配置文件创建路由对象，然后将路由配置文件引入 main.js 入口文件并注册到 Vue 实例上。上面的流程操作完成后，就可以在页面组件中使用路由的内置组件 router-link 和 router-view 进行页面跳转了。

/router/index.js 文件代码如下：

```
import Vue from 'vue'
import VueRouter from 'vue - router'

Vue.use(VueRouter)

const routes = [
  {
    path: '/home',
    name: 'Home',
    component:() => import('@/views/Home.vue')
  }
]

const router = new VueRouter({
  mode: 'history',
  base:process.env.BASE_URL,
  routes
})

export default router
```

/main.js 文件代码如下：

```
import Vue from 'vue'
import App from './App.vue'
import router from'./router'

new Vue({
```

```
  router,
  render: h => h(App)
}).$mount('#app')
```

/App.vue 文件代码如下：

```
<template>
  <div>
    <div>
      <!-- 用于跳转路由的链接,to 属性为跳转地址 -->
      <router-link to="/home">Home 页面</router-link>
    </div>

    <!-- 路由匹配的组件会渲染到 router-view -->
    <router-view/>
  </div>
</template>
```

/views/Home.vue 文件代码如下：

```
<template>
  <div>
    这是 Home 页面。
  </div>
</template>
```

在浏览器中运行，打开项目根目录会显示 router-link 的超链接效果，如图 6.1 所示。单击超链接后，跳转到 http://localhost:8080/home 路由下并渲染 Home.vue 视图，效果如图 6.2 所示。

图 6.1　项目的首页

图 6.2　Home.vue 页面

6.1.3　动态路由

很多时候，我们需要从一个页面跳转到另一个页面，并且携带参数，在这种应用场景下就可以使用动态路由。动态路由可以将某种模式匹配到所有路由，全部映射到同一个组件上。例如，我们需要访问一个商品页面的组件 goods.vue 文件，对于所有要访问这个页面组件的用户来说，都要使用这个组件进行视图渲染。那么就可以在 vue-router 的路由路径中使用“动态路径参数”来达到这个效果。

一个“路径参数”使用冒号:标记。当匹配到一个路由时，参数值会被设置到

this.$route.params，这样便可以在每个组件内使用。

/router/index.js 文件代码如下：

```
import Vue from 'vue'
import VueRouter from 'vue-router'

Vue.use(VueRouter)

const routes = [
  {
    path: '/goods/:gid',
    name: 'Goods',
    component:() => import('@/views/Goods.vue')
  }
]

const router = new VueRouter({
  mode: 'history',
  base:process.env.BASE_URL,
  routes
})

export default router
```

/App.vue 文件代码如下：

```
<template>
  <div>
    <div>
    <!-- 用于跳转路由的链接,to 属性为跳转地址 -->
      <router-link to="/goods/1001">查看商品</router-link>
    </div>

    <!-- 路由匹配的组件会渲染到 router-view -->
    <router-view/>
  </div>
</template>
```

/views/Goods.vue 文件代码如下：

```
<template>
    <div>
        商品详情页面
        <p>
            商品 ID: {{ $route.params.gid}}
        </p>
    </div>
</template>
```

在浏览器中运行，项目根目录下会显示“查看商品”的超链接，效果如图 6.3 所示。单击超链接，页面跳转到 /goods 商品详情路由下，并渲染 Goods. vue 视图，在商品详情页面中会显示传递过来的商品 ID 参数值，效果如图 6.4 所示。

图 6.3　项目首页效果

图 6.4　商品详情页效果

可以在一个路由中设置多段“路径参数”，对应的值都会设置到 $route. params 中。参数设置参考表 6.1。

表 6.1　路由路径设置参数

模式	匹配路径	$route. params
/goods/:gid	/goods/1001	{gid: '1001'}
/goods/:type/:gid	/goods/shoes/1001	{type:'shoes', gid: '1001'}

除了 $route. params 外，$route 对象还提供了其他有用的信息，例如，$route. query（在 URL 中设置查询参数）、$route. hash 等。

6.1.4　嵌套模式路由

实际生活中的应用界面通常由多层嵌套的组件组合而成，在配置路由的过程中，需要对 URL 进行分层管理，使每个路由都能按照嵌套的顺序进行编写。我们还是以商城类应用为例，在商品分类页面，单击某一个类别，要跳转到商品的列表页面，那么该商品列表页面的路由就由商品分类＋商品列表组成。

/App. vue 文件代码如下：

```
<template>
  <div>
    <!-- 路由匹配的组件会渲染到 router - view -->
    <router - view/>
  </div>
</template>
```

在上面的代码中 <router-view> 是最顶层的出口，渲染最高级路由匹配到的组件。同样地，一个被渲染组件同样可以包含自己的嵌套 <router-view>。例如，在 Classify. vue 组件的模板添加一个 <router-view>。

/views/Classify. vue 文件代码如下：

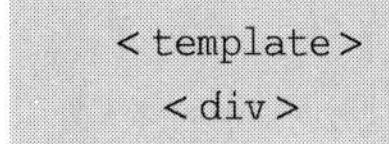

```
<template>
  <div>
```

```
    <div>
      <router-link to="/classify/list/1">男装</router-link>   
      <router-link to="/classify/list/2">女装</router-link>   
      <router-link to="/classify/list/3">童装</router-link>
    </div>

    <router-view></router-view>
  </div>
</template>
```

/views/GoodsList.vue 文件代码如下：

```
<template>
    <div>
        商品列表页 -- 分类 id: {{ $route.params.tid}}
    </div>
</template>
```

要在嵌套的出口中渲染组件，需要在 VueRouter 的参数中使用 children 配置。/router/index.js 文件代码如下：

```
import Vue from 'vue'
import VueRouter from 'vue-router'

Vue.use(VueRouter)

const routes = [
  {
    path: '/classify',
    name: 'Classify',
    component:() => import('@/views/Classify.vue'),
      children: [
        {
            path: '/classify/list/:tid',
            name: 'GoodsList',
            component: () => import('@/views/GoodsList.vue')
        }
      ]
  }
]

const router = new VueRouter({
  mode: 'history',
  base:process.env.BASE_URL,
  routes
})

export default router
```

在浏览器中运行,访问 /classify 路由时打开商品分类页面,效果如图 6.5 所示。当单击不同的商品类别时,跳转到 /classify/list/:tid 路由,渲染商品列表页面视图,并动态地获取当前商品列表的 id,效果如图 6.6 所示。

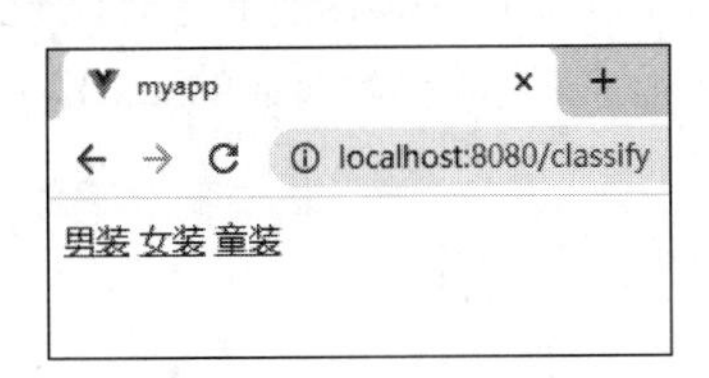

图 6.5　商品分类页面

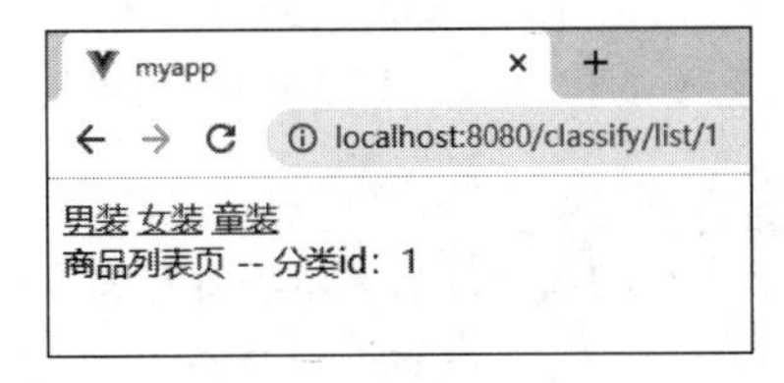

图 6.6　商品列表页面

6.1.5　编程式导航

除了使用 <router-link> 创建 a 标签来定义导航链接,还可以借助 router 的实例方法通过编写代码实现导航。

1. push 函数

语法:

```
router.push(location, onComplete?, onAbort?)
```

如果想要导航到不同的 URL,则可以使用 router.push 方法。这种方法会向 history 栈添加一个新的记录,所以当用户单击浏览器中的后退按钮时,则回到之前的 URL。在 Vue 实例内部,可以通过 $router 访问路由实例,具体操作为 this.$router.push。

当单击 <router-link> 时,这种方法会在内部调用,所以说,单击 <router-link :to="..."> 等同于调用 router.push(...)。

声明式代码如下:

```
<router-link :to="...">
```

编程式代码如下:

```
router.push(...)
```

该方法的参数可以是一个字符串路径,也可以是一个描述地址的对象,代码如下:

```
//字符串
router.push('home')

//对象
router.push({ path: 'home' })

//命名的路由
router.push({ name: 'user', params: { userId: '123' }})
```

```
//带查询参数,变成 /register?plan = private
router.push({ path: 'register', query: { plan: 'private' }})
```


如果提供了 path,params 会被忽略,上述例子中的 query 并不属于这种情况。取而代之的是下面例子的做法,需要提供路由的 name 或手写完整的带有参数的 path,代码如下:

```
const userId = '123'
router.push({ name: 'user', params: { userId }}) // -> /user/123
router.push({ path: '/user/${userId}' }) // -> /user/123
//这里的 params 不生效
router.push({ path: '/user', params: { userId }}) // -> /user
```

同样的规则也适用于 router-link 组件的 to 属性。

2. replace 函数

语法:

```
router.replace(location, onComplete?, onAbort?)
```

router.replace 与 router.push 很像,唯一的不同就是,它不会向 history 添加新记录,而是像它的方法名一样,替换掉当前的 history 记录。

声明式示例代码如下:

```
<router-link :to="..." replace>
```

编程式示例代码如下:

```
router.replace(...)
```

3. go 函数

语法:

```
router.go(n)
```

这种方法的参数是一个整数,意思是在 history 记录中向前或者后退多少步,类似 window.history.go(n),代码如下:

```
//在浏览器记录中前进 1 步,等同于 history.forward()
router.go(1)

//后退 1 步记录,等同于 history.back()
router.go(-1)

//前进 3 步记录
router.go(3)
```

```
//如果 history 记录不够用,那就默认失败
router.go(-100)
router.go(100)
```

6.2 路由的相关配置

6.2.1 命名路由

在开发过程中,如果每次使用路由跳转的过程都用 path 会比较麻烦,如果能通过一个名称来标识一个路由,则会更加方便。在 vue-router 中就有关于命名路由的配置项,创建 Router 实例的时候,在 routes 配置中可以给某个路由设置名称,代码如下:

```
const router = new VueRouter({
  routes: [
    {
      path: '/user/:userId',
      name: 'user',
      component: User
    }
  ]
})
```

要连接到一个命名路由,可以给 router-link 的 to 属性传一个对象,代码如下:

```
<router-link :to="{
          name: 'user',
          params: { userId: 123 }
      }">User</router-link>
```

使用代码调用 router.push()的效果是一样的,代码如下:

```
router.push({ name: 'user', params: { userId: 123 }})
```

上面这两种方式都可以把路由导航到 /user/123 路径。

6.2.2 命名视图

有时候想同时(同级)展示多个视图,而不是嵌套展示,例如创建一个布局,有 sidebar(侧导航)和 main(主内容)两个视图,这个时候命名视图就派上用场了。可以在界面中拥有多个单独命名的视图,而不是只有一个单独的出口。如果 router-view 没有设置名字,则默认为 default。

```
<router-view class="view one"></router-view>
<router-view class="view two" name="a"></router-view>
<router-view class="view three" name="b"></router-view>
```

一个视图使用一个组件渲染，因此对于同一个路由，多个视图就需要多个组件。确保正确使用 components 配置(带上 s)，代码如下：

```
const router = new VueRouter({
  routes: [
    {
      path: '/',
      components: {
      default: Foo,
        a: Bar,
        b: Baz
      }
    }
  ]
})
```

6.2.3 重定向

在实际开发中，当对一个页面的功能操作完成后，需要自动完成跳转，或者在访问某个路由链接时，需要自动访问另外一个链接，这就要用到路由的重定向配置。重定向可以通过 routes 配置来完成，代码如下：

```
const router = new VueRouter({
  routes: [
    { path: '/a', redirect: '/b'}
  ]
})
```

重定向的目标也可以是一个命名的路由，代码如下：

```
const router = new VueRouter({
  routes: [
    { path: '/a', redirect: { name: 'foo'}}
  ]
})
```

还可以通过一种方法，动态返回重定向目标，代码如下：

```
const router = new VueRouter({
  routes: [
    { path: '/a', redirect: to => {
      //方法接收目标路由作为参数
      //return 重定向的字符串路径/路径对象
    }}
  ]
})
```

6.3 路由的模式

在讲解 vue-router 的路由模式之前，首先要认识路由的组成。每个路由都是由多个 URL 组成，使用不同的 URL 可以导航到不同的位置。对于服务器端访问来说，HTTP 请求是无状态的，所以当请求服务器不同的网址来切换页面时，都会重新进行请求。而在使用 vue-router 进行前端页面切换时，并没有让浏览器刷新，这是因为借助了浏览器的 history API，使得页面跳转而浏览器不执行刷新操作，这样页面的状态就被维持在浏览器中了。

vue-router 中默认为哈希模式，URL 网址的格式为 http://localhost:8080/#/，在 URL 中带有#号。可以在 router 实例中修改路由的模式，代码如下：

```
const router = new VueRouter({
  mode: 'history',
  base:process.env.BASE_URL,
  routes
})
```

当路由的模式设置为 history 模式后，URL 网址中的#就会被去除了。

6.4 导航守卫

导航守卫又称为路由守卫，用来实时监控路由跳转的过程，在路由跳转的各个过程中执行相应的钩子函数，这就类似于 Vue 的生命周期钩子，在实际开发中经常被使用。例如，当用户单击一个页面时，如果当前用户未登录，就自动跳转到登录页面；如果已经登录，就让用户正常进入。

导航守卫分为全局守卫、路由独享守卫和组件内守卫，这 3 种方式应用的场景不同，都有自己的钩子函数，具体内容如下。

6.4.1 全局守卫

全局守卫的钩子函数有 3 个，分别是：

- router.beforeEach(全局前置守卫)
- router.beforeResolve(全局解析守卫)
- router.afterEach(全局后置守卫)

1. router.beforeEach(全局前置守卫)

可以使用 router.beforeEach 注册一个全局前置守卫，代码如下：

```
const router = new VueRouter({ ... })

router.beforeEach((to, from, next) => {
  //...
})
```

当一个导航触发时，全局前置守卫按照创建顺序调用。守卫的执行方式为异步解析执行，此时导航在所有守卫 resolve 完之前一直处于等待中。

每个守卫方法接收 3 个参数：

(1) to：是一个 Route 对象，表示即将进入的目标路由对象。

(2) from：是一个 Route 对象，表示当前导航正要离开的路由。

(3) next：是一个函数对象，必须调用该方法来完成这个钩子，执行效果依赖 next 方法的调用参数。

2. router.beforeResolve(全局解析守卫)

和全局前置守卫类似，其区别是在跳转被确认之前，同时在所有组件内守卫和异步路由组件都被解析之后，解析守卫才调用。

3. router.afterEach(全局后置钩子)

可以注册全局后置钩子，然而和守卫不同的是，这些钩子只接收 to 和 from 参数，不会接收 next 函数也不会改变导航本身。

6.4.2 路由独享守卫

独享守卫只有一种：beforeEnter。该守卫接收的参数与全局守卫是一样的，但是该守卫只在其他路由跳转至配置有 beforeEnter 路由表信息时才生效。

router 配置文件的配置如下：

```
const router = new VueRouter({
  routes: [
    {
      path: '/foo',
      component: Foo,
      beforeEnter: (to, from, next) => {
        //...
      }
    }
  ]
})
```

6.4.3 组件内守卫

组件内守卫是在组件内部直接定义的，有以下 3 个钩子函数。

(1) beforeRouteEnter：进入该路由前执行。

(2) beforeRouteUpdate：该路由的动态参数值发生改变时执行。

(3) beforeRouteLeave：离开该路由时执行。

代码如下：

```
const Foo = {
  template: '...',
  beforeRouteEnter (to, from, next) {
```

```
    //在渲染该组件的对应路由被加载前调用
    //不能获取组件实例'this'
    //因为当守卫执行前,组件实例还没被创建
  },
  beforeRouteUpdate (to, from, next) {
    //在当前路由改变,但是该组件被复用时调用
    //举例来说,对于一个带有动态参数的路径 /foo/:id,在 /foo/1 和 /foo/2 之间跳转的时候
    //由于会渲染同样的 Foo 组件,因此组件实例会被复用。而这个钩子就会在这个情况下被调用
    //可以访问组件实例'this'
  },
  beforeRouteLeave (to, from, next) {
    //导航离开该组件的对应路由时调用
    //可以访问组件实例'this'
  }
}
```

beforeRouteEnter 守卫不能访问 this,因为守卫在导航确认前被调用,因此即将登场的新组件还没被创建。

不过,可以通过传一个回调给 next 访问组件实例。在导航被确认的时候执行回调,并且把组件实例作为回调方法的参数,代码如下:

```
beforeRouteEnter (to, from, next) {
next(vm => {
    //通过 vm 访问组件实例
  })
}
```

注意　beforeRouteEnter 是支持向 next 传递回调的唯一守卫。对于 beforeRouteUpdate 和 beforeRouteLeave 来说,this 已经可用了,所以不支持传递回调,因为没有必要。

第7章 Vuex状态管理

7.1 Vuex简介

7.1.1 什么是Vuex

Vuex是一个专为Vue.js应用程序开发的状态管理工具。它采用了集中式存储管理应用的所有状态,并以相应的规则保证状态以一种可预测的方式发生变化。

简单来说,Vuex是一个适用于在Vue项目开发时使用的状态管理工具,如果在一个项目开发过程中频繁地使用组件传参的方式实现数据的同步,那么在项目的扩展、管理和维护方面将是一个灾难。为此,Vue为这些被多个组件频繁使用的数据提供了一个统一的管理工具,即Vuex。

在具有Vuex的Vue项目中,我们只需把这些值定义在Vuex的状态管理对象中,就可以在整个项目的组件内使用。

7.1.2 Vuex的安装与使用

在使用Vue CLI脚手架工具创建项目的过程中,可以手动选择安装Vuex,也可以在项目创建完成后再独立安装。

使用NPM安装Vuex命令如下:

```
npm i vuex -- save
```

使用yarn安装Vuex命令如下:

```
yarn add vuex
```

在一个模块化的打包系统中,必须显式地通过Vue.use()来注册Vuex。

/src/store/index.js文件代码如下:

```
import Vue from 'vue'
import Vuex from 'vuex'

//挂载Vuex
Vue.use(Vuex)
```

```
//创建 Vuex 对象
const store = new Vuex.Store({
    state:{
        //存放的键值对就是所要管理的状态
        name:'helloVuex'
    }
})

export default store
```

将 store 挂载到当前项目的 Vue 实例中。

/main.js 文件代码如下：

```
import Vue from 'vue'
import App from './App'
import router from './router'
import store from './store'

Vue.config.productionTip = false

/* eslint-disable no-new */
new Vue({
  el: '#app',
  router,
  store, //store:store 和 router 相同,将我们创建的 Vuex 实例挂载到这个 Vue 实例中
  render: h => h(App)
})
```

在组件中使用 Vuex。例如在 App.vue 中，我们要将 state 中定义的 name 在 h1 标签中显示。

/App.vue 文件代码如下：

```
<template>
    <div id='app'>
        name:
        <h1>{{ $store.state.name }}</h1>
    </div>
</template>
```

或者在组件方法中使用，代码如下：

```
//...
methods:{
    add(){
      console.log(this.$store.state.name)
    }
},
//...
```

7.2 Vuex 核心概念

7.2.1 Vuex 的工作流程

33min

在 Vuex 对象中定义了 5 个成员来管理对象内的数据，其中 state 用来存放数据，剩下的成员都用来操作 state 中数据的方法集。Vuex 对象的成员有以下 5 个。

（1）state：存放状态。

（2）mutations：state 成员操作。

（3）getters：加工 state 成员给外界。

（4）actions：异步操作。

（5）modules：模块化状态管理。

Vuex 的具体工作流程如图 7.1 所示。

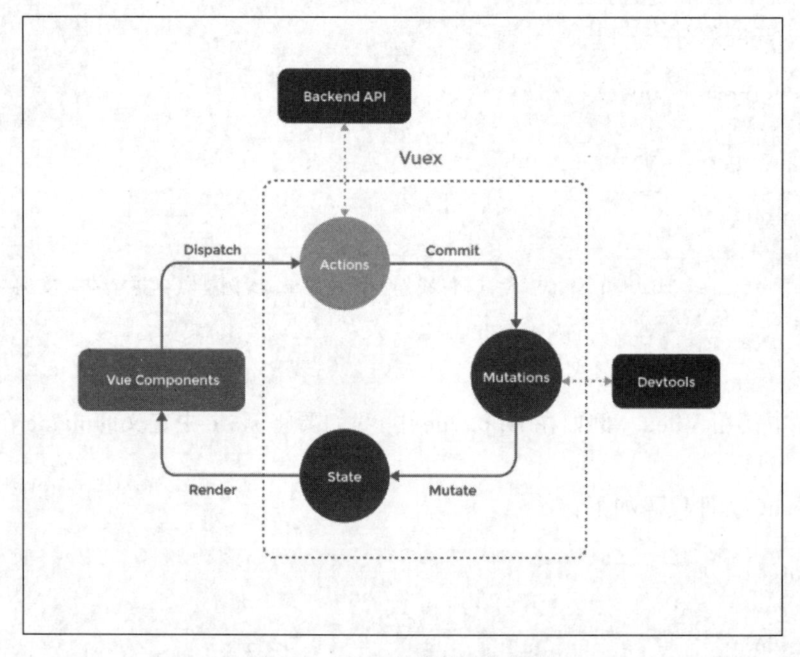

图 7.1　Vuex 的具体工作流程

在 Vuex 的整个工作流程中，会发现 actions 和 mutations 都是对 state 中的数据进行操作的方法集，但是它们被应用在不同的场景下。例如，Vue 中的组件如果在调用 Vuex 的某种方法时，需要向服务器发送请求或者需要进行其他的异步操作，这时就需要使用 $store.dispatch()将数据派发到 actions 的对应方法中，以此来保证数据的同步。actions 中方法的作用就是为了让 mutations 中的方法能在异步操作中起作用。

如果在操作 state 数据的过程中没有异步行为，那么就可以直接使用 $store.commit()将数据提交到 mutations 的方法中，然后在方法中对数据进行操作。更新后的 state 数据会被渲染到组件中。

7.2.2 Vuex 对象核心成员

在 Vuex 对象中一共有以下 5 个核心成员。

1. state

state 的用法非常简单，就是用来存放 Vuex 中的数据。

2. mutations

mutations 是操作 state 数据的方法的集合，例如对该数据的修改、增加、删除等。在 mutations 中定义的每种方法都有两个默认参数，语法如下：

```
method(state,payload){
}
```

形参 state 表示当前 Vuex 对象中的 state，形参 payload 是指该方法在被调用时传递的参数。

/store/index.js 文件代码如下：

```
import Vue from 'vue'
import Vuex from 'vuex'

Vue.use(Vuex)

export default new Vuex.Store({
  state: {
      num: 0
  },
  mutations: {
      add(state,i){
        state.num += i
      }
  }
})
```

在组件中使用 commit()方法提交数据，例如，在 App.vue 文件中调用 mutations 内定义的函数，代码如下：

```
this.$store.commit('add',10)
```

在上面的示例代码中，执行 $store.commit()方法可以将参数“10”传递给 mutations 内的 add()方法，并为 state 中的数据重新赋值。如果要传递多个参数，可以使用对象的数据类型进行传值，代码如下：

```
this.$store.commit('add',{i:10, j:11})
```

3. getters

可以对 state 中的成员做一系列运算后再向外返回数据，getters()中定义的方法有两个默认参数，代码如下：

```
import Vue from 'vue'
import Vuex from 'vuex'

Vue.use(Vuex)

export default new Vuex.Store({
  state: {
      name: '张三',
      age: 20
  },
  getters: {
      nameInfo(state){
          return "姓名:" + state.name
      },
      fullInfo(state,getters){
          return getters.nameInfo + " 年龄:" + state.age
      }
  }
})
```

getters()中定义的方法默认参数如下。

（1）state：当前 Vuex 对象中的状态对象。

（2）getters：当前 getters 对象，用于获取 getters 下的其他 getter。

在组件内调用 getters()方法，代码如下：

```
this.$store.getters.fullInfo
```

4. actions

由于直接在 mutations()方法中进行异步操作，将会引起数据失效。所以提供了 actions 来专门进行异步操作，最终提交 mutations()方法，代码如下：

```
import Vue from 'vue'
import Vuex from 'vuex'

Vue.use(Vuex)

export default new Vuex.Store({
  state: {
      num: 0
  },
  mutations: {
      add(state,payload){
          state.num += payload
      }
  }
  actions: {
```

```
        add(context,payload){
            setTimeout(() =>{
                context.commit('add',payload)
            },2000)
        }
    }
})
```

actions 中的方法有两个默认参数,分别如下。

(1) context:上下文(相当于箭头函数中的 this)对象。

(2) payload:挂载参数。

在组件内调用 actions 中定义的函数,代码如下:

```
this.$store.dispatch('add',10)
```

由于是异步操作,所以我们可以将 actions 中的异步方法封装为一个 Promise 对象,代码如下:

```
add(context,payload){
        return new Promise((resolve,reject) =>{
            setTimeout(() =>{
                context.commit('add',payload)
                resolve()
            },2000)
        })
}
```

5. modules

当项目庞大,并且状态非常多时,可以采用模块化管理模式。Vuex 允许将 store 分割成模块(module)。每个模块拥有自己的 state、mutations、actions、getters,甚至是嵌套子模块——从上至下进行同样方式分割,代码如下:

```
import Vue from 'vue'
import Vuex from 'vuex'

Vue.use(Vuex)

export default new Vuex.Store({
  modules: {
      a:{
          state: {},
          getters: {},
          //...
      }
  }
})
```

在组件内调用模块a的状态，代码如下：

```
this.$store.state.a
```

而提交或者dispatch某种方法和以前一样，会自动执行所有模块内对应的type()方法，代码如下：

```
this.$store.commit('editKey')
this.$store.dispatch('aEditKey')
```

在使用modules模块时，需要注意以下细节。

模块中mutations()和getters()方法接收的第1个参数是自身局部模块内部的state，代码如下：

```
models:{
    a:{
        state:{key:5},
        mutations:{
            editKey(state){
                state.key = 9
            }
        },
        …
    }
}
```

getters()方法中的第3个参数表示根节点状态，代码如下：

```
models:{
    a:{
        state:{key:5},
        getters:{
            getKeyCount(state,getter,rootState){
                return  rootState.key + state.key
            }
        },
        …
    }
}
```

actions中方法获取局部模块状态的命令为context.state，获取根节点状态的命令为context.rootState，代码如下：

```
models:{
    a:{
        state:{key:5},
        actions:{
```

```
            aEidtKey(context){
                if(context.state.key === context.rootState.key){
                    context.commit('editKey')
                }
            }
        },
        ...
    }
}
```

7.2.3 Vuex 规范目录结构

因为把整个 store 放到 index.js 中是不合理的，所以需要拆分。示例目录格式如下：

```
|- store
  |-  actions.js
  |-  getters.js
  |-  index.js
  |-  mutations.js
  |-  mutations_type.js   ##该项为存放 mutaions 方法常量的文件，按需要可加入
  |
  |- modules
    |- Astore.js
```

对应的内容存放在对应的文件中，在/store/index.js 文件中存放并导出 store 对象。state 中的数据尽量放到 index.js 中。而 modules 中的 Astore 局部模块状态如果多也可以进行细分。

第8章 Vue的异步请求

35min

8.1 axios的安装与使用

axios是一个基于promise的HTTP库，主要用来向服务器端发起请求，可以在请求中做更多的可控操作，例如拦截请求等。

axios可以在浏览器和Node.js中使用，Vue、React等前端框架的广泛普及，促使了axios这种轻量级库的出现。

axios的特性：

（1）可以在浏览器中发送XMLHttpRequests。

（2）可以在Node.js中发送HTTP请求。

（3）支持Promise API。

（4）拦截请求和响应。

（5）转换请求数据和响应数据。

（6）能够取消请求。

（7）自动转换JSON数据。

（8）客户端支持保护安全，免受XSRF攻击。

8.1.1 安装axios

安装命令如下：

```
npm install axios --save
```

打开使用Vue脚手架创建的项目，在main.js文件中引入axios模块，代码如下：

```
import axios from 'axios';
Vue.prototype.$axios = axios;
```

在组件中使用axios发送异步请求，代码如下：

```
<script>
    export default {
        mounted(){
            this.$axios.get('/user?id=123').then(res =>{
```

```
                    console.log(res.data);
                })
            }
        }
    </script>
```

8.1.2 axios 基本用法

axios 主要的作用是向服务器端发起 HTTP 请求，根据 HTTP 标准，HTTP 请求可以使用多种请求方法。

为了在开发中能够更方便地使用 axios，axios 为所有支持的请求方法提供了别名。

- axios.request(config)
- axios.get(url[,config])
- axios.delete(url[,config])
- axios.head(url[,config])
- axios.options(url[,config])
- axios.post(url[,data[,config]])
- axios.put(url[,data[,config]])
- axios.patch(url[,data[,config]])

注意 在使用别名方法时，url、method、data 这些属性都不必在配置中指定。

1. GET 请求

GET 请求用于获取数据，从指定的资源请求数据，并返回实体主体，代码如下：

```
//方法一
this.$axios.get('/URL',{
    params: {
        id:1
    }
}).then(res =>{
        console.log(res.data);
    },err =>{
        console.log(err);
})

//方法二
this.$axios({
    method: 'get',
    url: '/URL',
    params: {
        id:1
    }
}).then(res =>{
    console.log(res.data);
},err =>{
    console.log(err);
})
```

2. POST 请求

POST 请求是向指定资源提交数据并处理请求(例如提交表单或者上传文件)。数据被包含在请求体中。

POST 请求一般分为两种类型:

(1) form-data 表单提交,图片上传、文件上传时使用该类型比较多。

(2) application/json 一般是用于 ajax 异步请求。

代码如下:

```
this.$axios.post('/URL',{
    id:1
}).then(res =>{
    console.log(res.data);
},err =>{
    console.log(err);
})
```

3. PUT 请求

PUT 请求用于更新数据,从客户端向服务器传送的数据取代指定的文档内容,代码如下:

```
this.$axios.put('/URL',{
    id:1
}).then(res =>{
    console.log(res.data);
})
```

4. PATCH 请求

PATCH 请求也被用于更新数据,是对 put 方法的补充,用来对已知资源进行局部更新,代码如下:

```
//PATCH 请求
this.$axios.patch('/URL',{
    id:1
}).then(res =>{
    console.log(res.data);
})
```

5. DELETE 请求

DELETE 请求服务器删除指定的页面。使用 axios 发送 DELETE 请求,参数可以使用明文的方式或者封装对象的方式进行提交,代码如下:

```
//参数以明文方式提交
this.$axios.delete('/URL',{
    params: {
        id:1
```

```
    }
}).then(res =>{
    console.log(res.data);
})

//参数以封装对象方式提交
this.$axios.delete('/URL',{
    data: {
        id:1
    }
}).then(res =>{
    console.log(res.data);
})
```

8.2 axios 实例

当 axios 要请求多个不同的后端接口地址，并且一些 axios 配置项都相同时，可以先创建 axios 实例，然后使用 axios 实例发起请求。

可以使用自定义配置新建一个 axios 实例，代码如下：

```
let instance = this.$axios.create({
    baseurl: 'http://localhost:9090',
        timeout:2000
});

instance.get('/query').then(res =>{
    console.log(res.data);
});
```

在一个 Vue 组件中可以同时创建多个 axios 实例。

axios 实例常用配置：

(1) baseurl：请求的域名基本网址，类型为 String。

(2) timeout：请求超时时长，单位为 ms，类型为 Number。

(3) URL：请求路径，类型为 String。

(4) method：请求方法，类型为 String。

(5) headers：设置请求头，类型为 Object。

(6) params：请求参数，将参数拼接在 URL 上，类型为 Object。

(7) data：请求参数，将参数放到请求体中，类型为 Object。

1. axios 全局配置

代码如下：

```
//配置全局的超时时长
this.$axios.defaults.timeout = 2000;
```

```
//配置全局的基本 URL
this.$axios.defaults.baseurl = 'http://localhost:8080';
```

2. axios 实例配置

代码如下：

```
let instance = this.$axios.create();
instance.defaults.timeout = 3000;
```

3. axios 请求配置

代码如下：

```
this.$axios.get('/query',{
    timeout: 3000
}).then();
```

以上配置的优先级为请求配置→实例配置→全局配置。

8.3 axios 并发请求

axios 提供了并发请求的方法，可以同时进行多个请求，并统一处理返回值，代码如下：

```
this.$axios.all([
    this.$axios.get('/query/classify'),
    this.$axios.get('/query/goods')
]).then(
    this.$axios.spread((classifyRes,goodsRes) =>{
        console.log(classifyRes.data);
        console.log(goodsRes.data);
    })
)
```

8.4 axios 拦截器

axios 提供了拦截器功能，使用拦截器可以提高请求的可控性，并且完成更多复杂的操作。axios 的拦截器分为请求拦截器和响应拦截器，两种拦截器在不同的时机对 axios 发起的请求进行处理。

1. 请求拦截器

在请求被 then 或 catch 处理前拦截它们，代码如下：

```
this.$axios.interceptors.request.use(config =>{
    //发生请求前处理
    return config;
```

```
},err=>{
    //请求错误处理
    return Promise.reject(err);
});

//或者用 axios 实例创建拦截器
let instance =  $axios.create();
instance.interceptors.request.use(config=>{
    return config
});
```

2. 响应拦截器

在响应被 then 或 catch 处理前拦截它们，代码如下：

```
this.$axios.interceptors.response.use(res=>{
    //请求成功对响应数据进行处理
    //该返回对象会传到请求方法的响应对象中
    return res
},err=>{
    //响应错误处理
    return Promise.reject(err);
});
```

3. 取消拦截

如果想在稍后移除拦截器，则可以设置取消拦截，代码如下：

```
let instance = this.$axios.interceptors.request.use(config=>{/*...*/});

//取消拦截
this.$axios.interceptors.request.eject(instance);
```

8.5 axios 错误处理

axios 请求拦截器和响应拦截器抛出错误时，返回的 err 对象会传给 catch()函数的 err 对象参数，代码如下：

```
this.$axios.get('/URL').then(res=>{
    //获取响应数据
}).catch(err=>{
    //错误处理
    console.log(err);
});
```

8.6 axios 取消请求处理

axios 取消请求主要用于取消正在进行的 HTTP 请求，代码如下：

```
let source = this.$axios.CancelToken.source();

this.$axios.get('/goods.json',{
    cancelToken: source.token
}).then(res =>{
    console.log(res)
}).catch(err =>{
    //取消请求后会执行该方法
    console.log(err)
});

//取消请求,参数可选,该参数信息会发送到请求的 catch 中
source.cancel('取消后的信息');
```

第9章 服务器端渲染

9.1 服务器端渲染简介

9.1.1 什么是服务器端渲染(SSR)

服务器端渲染(Server Side Render,SSR)。Vue.js 用于构建客户端应用程序的框架,在默认情况下,在浏览器中输出 Vue 组件,生成 DOM 和操作 DOM。但是这种操作对 SEO 不利,所以在开发过程中需要在服务器端将组件渲染为 HTML 字符串,然后将它们直接发送到浏览器端。

简单来说,服务器端渲染就是将本来要放在浏览器进行创建的标签,放到服务器端先创建好,然后生成对应的 HTML 内容并直接发送到浏览器,最后将这些静态标记"激活"为客户端完全可交互的应用程序。

9.1.2 为什么要使用服务器端渲染

与传统的单页面应用程序相比,服务器端渲染的优势主要有以下几个方面。

(1) 更好的 SEO,让搜索引擎和爬虫抓取工具可以直接查看完全渲染的页面。

(2) 更快的内容到达时间,特别是对于缓慢的网络情况或运行缓慢的设备。

在使用服务器端渲染技术开发的网站中,也是伴随着一些缺点,需要权衡的方面有以下几点。

(1) 由于开发条件的限制,对于一些外部扩展库有时需要进行特殊处理,这样才能在服务器渲染应用程序中运行。

(2) 涉及构建设置和部署的更多要求,服务器端渲染的应用程序需要处于 Node.js server 运行环境。

(3) 更多的服务器端负载,在 Node.js 中渲染完整的应用程序会占用大量 CPU 资源和流量资源。

在做技术选型时,到底要不要使用服务器端渲染应该取决于项目的实际需求,这主要取决于内容到达时间对应用程序的重要程度。

如图 9.1 和图 9.2 所示,可以看到单页面应用网站与服务器端渲染网站的区别,使用 Google Chrome 浏览器分别打开两个网站,并右键查看源码。

可以明显看出服务器端渲染的源码量增加了好多倍。简单来说,服务器端渲染的模式就是,浏览器在请求一个网址的时候,服务器端接收到请求后把生成的 HTML 字符串响应给浏览器，当搜索引擎或爬虫在抓取页面内容时，可以轻松地获取 <a> 标签及其指向的网址。以此类推，搜索引擎就可以收录网站中的所有路径了。

```
<!DOCTYPE html><html><head><meta charset=utf-8><meta content="width=device-width,initial-scale=0.35,maximum-scale=2,minimum-scale=0.35," name=viewport><meta name=msapplication-tap-highlight content=no><meta name=baidu-site-verification content=Xbqfb2ndUt><meta content="telephone=no" name=format-detection><meta content="email=no" name=format-detection><meta name=apple-mobile-web-app-capable content=yes><meta name=apple-mobile-web-app-status-bar-style content=black-translucent><meta name=apple-mobile-web-app-title content=年糕妈妈><script type=text/javascript src="http://api.map.baidu.com/api?v=2.0&ak=BHllUXOTkwF4dGjdsgii8eIvIkgdX4qe"></script><title>年糕妈妈官网—让中国宝宝得到更科学的养育</title><meta name=keywords content=年糕妈妈,育儿网,年糕妈妈优选商城,母婴公众号,年糕妈妈公众号><meta name=description content=年糕妈妈, 中国领先的母婴知识服务及电商公司。旗下包含年糕妈妈、年糕妈妈学院、年糕妈妈教育、年糕妈妈辅食、陪玩时光等微信公号、糕妈优选电商平台以及nicomama定制等自有品牌的产品><link rel="shortcut icon" href=./static/favicon.ico><script type=text/javascript>var _vds = _vds || [];
    window._vds = _vds;
    (function() {
        _vds.push(['setAccountId', 'bc3404afb21f8505']);
        (function() {
            var vds = document.createElement('script');
            vds.type = 'text/javascript';
            vds.async = true;
            vds.src = ('https:' == document.location.protocol ? 'https://' : 'http://') + 'dn-growing.qbox.me/vds.js';
            var s = document.getElementsByTagName('script')[0];
            s.parentNode.insertBefore(vds, s);
        })();
    })();</script><script>var _hmt = _hmt || [];
    (function() {
      var hm = document.createElement("script");
      hm.src = "https://hm.baidu.com/hm.js?8547ee961679cd4d8bda494d2f1b5a03";
      var s = document.getElementsByTagName("script")[0];
      s.parentNode.insertBefore(hm, s);
    })();</script><link href=/static/css/1539676256320-app.00fb06e596df8f4189a9ac25b532b735.css rel=stylesheet></head><body><div id=app></div><script type=text/javascript src=/static/js/1539676256320-manifest.c127d140046b77b5741f.js></script><script type=text/javascript src=/static/js/1539676256320-vendor.094f02457de361c05de8.js></script><script type=text/javascript src=/static/js/1539676256320-app.9c5d198c7976864df240.js></script></body></html>
```

图 9.1　单页面应用网站源码

图 9.2　服务器端渲染网站源码

9.2　服务器端渲染的基本用法

9.2.1　安装与使用

1. 安装 vue-server-renderer

打开命令行工具，执行下面的命令，安装 vue-server-renderer 工具。命令如下：

```
npm install vue vue - server - renderer -- save
```

在安装 vue-server-renderer 工具时需要注意几个相关的问题。

(1) 使用服务器端渲染的 Node.js 版本要大于或等于 6.0 版本。

(2) vue-server-renderer 和 Vue 必须匹配版本。

(3) vue-server-renderer 依赖一些 Node.js 原生模块,因此只能在 Node.js 中使用。

2. 渲染一个 Vue 实例

代码如下:

```
//第 1 步: 创建一个 Vue 实例
const Vue = require('vue')
const app = new Vue({
  template: '<div>Hello World</div>'
})

//第 2 步: 创建一个 renderer
const renderer = require('vue-server-renderer').createRenderer()

//第 3 步: 将 Vue 实例渲染为 HTML
renderer.renderToString(app, (err, html) => {
  if (err) throw err
  console.log(html)
  // => <div data-server-rendered="true">Hello World</div>
})

//在 2.5.0+,如果没有传入回调函数,则会返回 Promise
renderer.renderToString(app).then(html => {
  console.log(html)
}).catch(err => {
  console.error(err)
})
```

9.2.2 与服务器集成

Vue.js 的服务器端渲染要依赖于 Node.js server 的运行环境,在实现的过程中可以直接使用 Node.js 的框架完成,例如 Express 框架。

在本地创建项目的目录,并在项目的根目录下打开命令行工具,安装 Express 框架。命令如下:

```
npm install express --save
```

在项目的根目录下创建 index.js 文件,代码如下:

```
const Vue = require('vue')
const server = require('express')()
const renderer = require('vue-server-renderer').createRenderer()

server.get('*', (req, res) => {
  const app = new Vue({
    data: {
```

```
      url: req.url
    },
    template: '<div>访问的 URL 是: {{ url }}</div>'
  })

  renderer.renderToString(app, (err, html) => {
    if (err) {
      res.status(500).end('Internal Server Error')
      return
    }
    res.end('
      <!DOCTYPE html>
      <html lang="en">
        <head><title>Hello</title></head>
        <body>${html}</body>
      </html>
    ')
  })
})

server.listen(8080)
```

9.3 Nuxt.js 框架

9.3.1 Nuxt.js 简介

Nuxt.js 是一个基于 Vue.js 的服务器端渲染应用框架，是由 zeit.co 幕后团队于 2016 年 10 月发布。Nuxt.js 预设了利用 Vue.js 开发服务器端渲染的应用所有需要的配置，只关注应用的 UI 渲染，可以在已有的 Node.js 项目中使用 Nuxt.js 框架。

Nuxt.js 继承了以下组件和框架，用于开发完整而强大的 Web 应用。

- Vue 2.x
- Vue-Router
- Vuex（当配置了 Vuex 状态树配置项时才会引入）
- Vue 服务器端渲染
- Vue-Meta

Nuxt.js 还使用了 Webpack、Vue-loader、babel-loader 来处理代码的自动化构建工作，例如打包、代码分层、压缩等操作。

我们可以使用 Nuxt.js 作为框架来处理项目的所有 UI 渲染，启动 Nuxt 时，它将启动具有热更新加载的开发服务器，并且 Vue 服务器端渲染配置可以自动为服务器呈现应用程序。

9.3.2 Nuxt.js 的项目搭建

在本地创建项目目录，在项目根目录下打开命令行工具，执行初始化命令。命令如下：

```
npm init -y
```

上面的命令执行成功后，会在项目根目录下自动创建 package.json 配置文件，然后在命令行工具中执行安装 nuxt 的命令。命令如下：

```
npm install nuxt -- save
```

上面的命令执行成功后，打开 package.json 文件，添加启动与 nuxt 相关的配置项，配置如下：

```
{
  "scripts": {
    "dev": "nuxt"
  }
}
```

上面的配置完成后，在项目的根目录下创建 pages 目录，Nuxt.js 会依据 pages 目录中的所有 *.vue 文件生成应用的路由配置。

在项目根目录下的命令行工具中执行下面的命令，创建 pages 目录。

```
mkdir pages
```

上面的命令执行成功后，进入 pages 目录，创建项目的第一个页面 pages/index.vue，代码如下：

```
<template>
  <h1>Hello world!</h1>
</template>
```

然后执行下面的命令启动项目。命令如下：

```
npm run dev
```

上面的命令执行成功后，在命令行会显示如图 9.3 所示的提示内容。

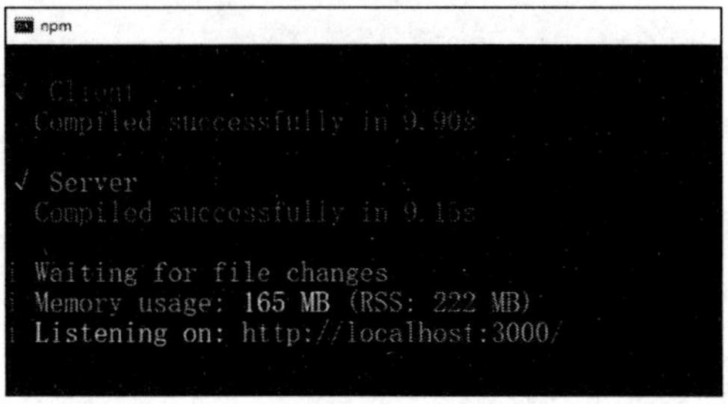

```
npm

√ Client
  Compiled successfully in 9.90s

√ Server
  Compiled successfully in 9.15s

i Waiting for file changes
i Memory usage: 165 MB (RSS: 222 MB)
i Listening on: http://localhost:3000/
```

图 9.3　项目启动成功提示

现在项目应用在 http://localhost:3000 上成功运行。在浏览器中访问项目网址，效果如图 9.4 所示。

图 9.4 在浏览器中运行项目效果

9.3.3 目录结构

Nuxt.js 的默认应用目录架构提供了良好的代码分层结构，适用于开发或大或小的应用。当然，也可以根据自己的偏好组织应用代码。

具体目录结构如下。

(1) 资源目录：资源目录 assets 用于组织未编译的静态资源，如 LESS、SASS 或 JavaScript。

(2) 组件目录：组件目录 components 用于组织应用的 Vue.js 组件。

(3) 布局目录：布局目录 layouts 用于组织应用的布局组件。

(4) 中间件目录：middlewares 目录用于存放应用的中间件。

(5) 页面目录：页面目录 pages 用于组织应用的路由及视图。

(6) 插件目录：插件目录 plugins 用于组织那些需要在根 Vue.js 应用实例化之前需要运行的 JavaScript 插件。

(7) 静态文件目录：静态文件目录 static 用于存放应用的静态文件，此类文件不会被 Nuxt.js 调用 Webpack 进行构建编译处理。服务器启动的时候，该目录下的文件会映射到应用的根路径/下。

(8) store 目录：store 目录用于组织应用的 Vuex 状态树文件。

(9) nuxt.config.js 文件：nuxt.config.js 文件用于组织 Nuxt.js 应用的个性化配置，以便覆盖默认配置。若无额外配置，该文件不能被重命名。

(10) package.json 文件：package.json 文件用于描述应用的依赖关系和对外暴露的脚本接口。该文件不能被重命名。

第10章 Vue 3 新特性详讲

10.1 为什么要用 Vue 3

在学习 Vue 3 的新特性之前，先来看一下 Vue 3 设计的目的是什么，为什么要对 Vue 2 做出很大的改变，以及 Vue 3 到底解决了什么问题。像 Vue 这样全球闻名的前端框架，在任何一次改动时，设计者都是经过深思熟虑的权衡，所以，Vue 3 的出现肯定是解决了某些棘手的问题。下面介绍一下 Vue 2 中遇到的问题。

10.1.1 Vue 2 对复杂功能的处理不友好

在使用 Vue 2 开发项目的过程中，随着更加复杂的业务逻辑的增加，复杂组件的代码变得难以维护。尤其是一个开发人员从别的开发人员手中接过一个新项目时，这个问题更突出。究其根本原因，Vue 2 中的 API 是通过选项来组织代码的，但是大部分情况下，通过逻辑来组织代码会更有意义。Vue 2 中缺少多个组件之间提取和复用逻辑的机制，现有的重用机制都有很多缺点。

假设有一个需求，要对所有的视频进行分类管理，单击不同的学科分类，可以查看不同的视频列表，如图 10.1 所示。针对这个需求，可以设置一个 filter（过滤）功能，根据用户单击的学科分类，展示不同内容。

图 10.1 视频分类效果

在上面的案例中，用户单击不同的学科分类，如果展示的视频列表超出了当前页面展示的数量，例如，每个分页下可展示 40 个视频，则超出范围的视频列表要使用分页显示，效果如图 10.2 所示。

图 10.2 视频列表分页效果

上面的案例，如果使用 Vue 2 开发，示例代码如下：

```
export default {
    data() {
        return {
            filter: {},                    //处理过滤功能
            pagination: {}                 //处理分页功能
        }
    },
    methods: {
        filterMethod: () => {},            //处理过滤功能
        paginationMethod: () => {}         //处理分页功能
    },
    computed: {
        …
    }
}
```

通过上面代码可以看到，在 data 中已经做了 filter（过滤）和 pagination（分页）的相关处理，但是在下面的 methods 和 computed 中还需要继续做相关的处理，这些功能会分散到好几个部分。如果仅仅是这两个处理功能，项目的逻辑看上去还不是特别复杂，但是如果再增加搜索、收藏、排序等功能，随着功能复杂度的上升，带来的问题也愈加明显。

10.1.2 Vue 2 中 mixin 存在缺陷

10.1.1 节中提到的案例，Vue 2 也给出了解决方案，那就是使用 mixin 混入的方式。对上面案例中遇到的问题，可以使用 mixin 重写编写代码，示例代码如下：

```
const filterMixin = {
    data() {
        return {}
    },
    methods: {}
```

```
}

const paginationMixin = {
    data() {
        return {}
    },
    methods: {}
}

export default {
    mixins: [filterMixin, paginationMixin]
}
```

在上面代码中，创建了两个 mixin 混入对象，filterMixin 和 paginationMixin，然后在 Vue 组件对象中使用 mixins 选项属性引入这两个对象。虽然这样可以暂时解决一些按逻辑分类的问题，但是这样做也会带来一些其他问题。

首先是会产生多个混入对象的属性和方法名称的命名冲突；其次是 mixin 对象所暴露的变量没有什么作用；然后是把混入对象中的逻辑复用到其他的组件中，还会出现一些不可预知的问题。

10.1.3 Vue 2 对 TypeScript 的支持有限

Vue 框架的开发者都清楚，Vue 2 对 TypeScript 的支持并不友好，这是因为在 Vue 中是依赖 this 上下文对象向外暴露属性，但是在组件中的 this 与普通的 JavaScript 中的 Object 对象处理的方式不同。其实，在 Vue 2 设计时就没有考虑对 TypeScript 的集成和强制类型的相关问题，所以才导致在 Vue 2 中使用 TypeScript 有很多阻碍。

10.2 Vue 3 简介

众所周知，前端技术一直更新得很快，特别是前端框架，更新速度更是极快的。在 2020 年 4 月 21 日晚上，Vue 的作者尤雨溪在 B 站上直播分享了 Vue 3 Beta 的最新进展，直到 9 月 19 日，Vue 3 正式版才发布。这个耗时两年，历经 99 位代码贡献者，2600 多次代码提交的大版本更新终于和众多开发者见面了。

为了减少前端开发者的学习成本，Vue 2 的大部分特性保留到了 Vue 3 中，开发者可以像使用 Vue 2 一样，原封不动地使用 Vue 3，这是遵循了渐进式的准则。如果你是一个保守派，只想使用 Vue 2 的写法，也是完全没有问题的。

Vue 3 增加了以下新特性。

1. Vue 3 在性能上有很大提升

但没有哪一位开发者不想要更快、更轻的框架。Vue 3 给开发者带来了极致的开发体验。整个 Vue 3 的代码库被重新编写成了一系列独立的、可实现不同功能的模块。据官方介绍，Vue 3 的代码打包大小减少了 41%，初次渲染速度提升了 55%，更新效率提升了 33%，内存使用率减少了 54%。这些数据都得益于 Vue 3 中重构了虚拟 DOM 的写法，提升渲染速度。

2. Vue 3 推出了新的 API

在 Vue 2 中遇到了一些问题，例如，复杂组件的代码变得越来越难以维护，缺少一种纯粹的多组件之间提取和复用逻辑的机制。虽然 Vue 2 中也提供了相关的解决方案，但是在 Vue 2 中对于重用机制这一部分也存在一些弊端。所以，Vue 3 中设计了 Composition API，这也是本章重点介绍和使用的 Vue 3 的新特性。

Composition 这个单词是“组合”的意思，是 Vue 3 新推出的一系列 API 的合集，主要包括了以下 API。

（1）ref。

（2）reactive。

（3）computed。

（4）watch。

（5）新的生命周期函数。

（6）支持自定义 Hooks 函数。

（7）Teleport。

（8）Suspense。

（9）全局 API 的修改和优化。

3. 更好地支持 TypeScript

有在 Vue 2 中集成 TypeScript 的开发者应该都体会过其中的痛苦，因为 Vue 2 在推出的时候没有把 TypeScript 作为一个考量范围，所以在设计 Vue 3 的时候，设计者们就痛定思痛，考虑了这方面的问题。

Vue 3 的源代码全部都是使用 TypeScript 语法编写的，提供了非常完备的类型定义，在使用 Vue 3 开发项目时，可以把 TypeScript 语法深入到各个大型项目中，让开发者更加方便地享受类型推论等一系列 TypeScript 的红利。同时，还可以在 VSCode 等编辑器中安装相关的插件，完美地使用 TypeScript 的各种功能。

10.3 Vue 3 项目搭建

10.3.1 Vue CLI 脚手架简介

学习 Vue 3 之前要先配置 Vue 3 和 TypeScript 的开发环境，本章节中使用的是 Vue 开发团队推出的官方脚手架工具——Vue CLI。

Vue CLI 是一个基于 Vue.js 进行快速开发的完整系统，它提供了一系列与 Vue 框架相关的功能，例如，启动一个本地服务器、静态校验代码格式、运行单元测试、构建生产环境等。

在安装 Vue CLI 脚手架工具之前，需要先检查一下 Node.js 的版本，推荐使用 Node.js 10 以上的版本。在 cmd 命令行工具中运行命令查看 Node.js 的版本，命令如下：

```
node --version
```

命令运行效果如图 10.3 所示。

```
C:\Windows\System32\cmd.exe

C:\Users\mac\Desktop\project>node --version
v12.13.1
```

图 10.3　查看 Node.js 版本

10.3.2　安装 Vue CLI

配置 Vue 3 和 TypeScript 的开发环境之前，要先安装 Vue CLI 脚手架工具，安装命令如下：

```
npm install -g @vue/cli
# OR
yarn global add @vue/cli
```

由于 npm 访问的是境外服务器，很多情况下会出现请求速度慢或请求不到服务器的问题，推荐使用 cnpm 的方式下载 Vue CLI，cnpm 安装的命令如下：

```
npm install -g cnpm --registry=https://registry.npm.taobao.org
```

使用 cnpm 安装 Vue CLI，命令如下：

```
cnpm install -g @vue/cli
```

如果之前安装过 Vue CLI，建议执行上面的命令来更新版本。安装成功后需要查看一下当前的 Vue CLI 版本是否为 4.x，命令如下：

```
vue --version
```

查看版本命令运行效果如图 10.4 所示。

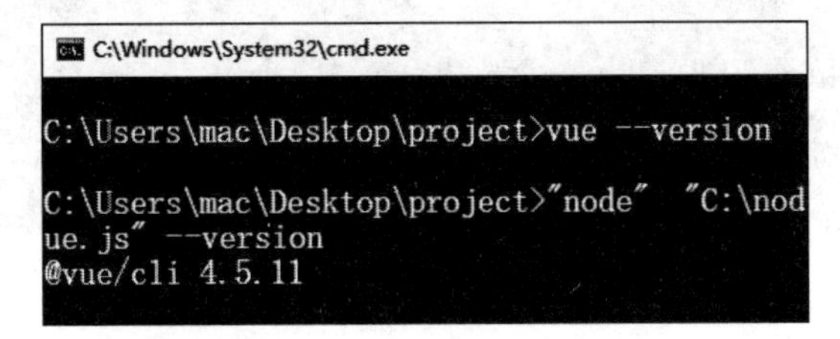

图 10.4　查看 Vue CLI 版本

本节中使用的 Vue CLI 的版本为 4.5.11，如果你的版本与本节中的版本不一致，不会造成负面影响，只需保证你的版本为 4.5.0 以上即可，因为只有 4.5.0 以上的版本才能创建支持最新版的 Vue 3 的基础项目。

10.3.3　创建 Vue 3 项目

Vue CLI 提供了命令行和 UI 界面创建项目的两种方式，无论使用哪种方式，创建的流程是完全一样的，只是展示的形式不太一样而已。本节使用命令行的方式创建 Vue 3 项目。

在本地硬盘中创建一个 project 的目录，在该目录下启动命令行工具，创建 vue3-basic 的项目，命令如下：

```
vue create vue3 - basic
```

在命令行工具中输入上面的命令，按回车键，会出现如图 10.5 所示的选项。

```
npm
Vue CLI v4.5.11
? Please pick a preset:
  Default ([Vue 2] babel, eslint)
  Default (Vue 3 Preview) ([Vue 3] babel, eslint)
> Manually select features
```

图 10.5　选择创建 Vue 项目模式

在图 10.5 中有 3 种创建模式，前两种分别是使用 Vue 2 和 Vue 3 默认的模板创建，第 3 种是手动创建，通过键盘的上下键选择创建模式，本节使用第 3 种手动创建的模式。

如果在你的命令行中没有出现 Vue 3 Preview 选项，说明现在使用的是旧版的 Vue CLI 脚手架工具，需要更新版本，更新版本的方法参考 10.3.2 节中的安装 Vue CLI 命令。

由于前两项创建模式不支持 TypeScript 语法，所以在命令行中，使用上下键选择 Manually select features 模式，按回车键进入下一步操作，效果如图 10.6 所示。

```
npm
Vue CLI v4.5.11
? Please pick a preset: Manually select features
? Check the features needed for your project: (Press <space> to select, <a> to toggle a
ll, <i> to invert selection)
>(*) Choose Vue version
 (*) Babel
 ( ) TypeScript
 ( ) Progressive Web App (PWA) Support
 ( ) Router
 ( ) Vuex
 ( ) CSS Pre-processors
 (*) Linter / Formatter
 ( ) Unit Testing
 ( ) E2E Testing
```

图 10.6　安装需要的模块

在图 10.6 中提供了一系列可插拔的支持，涵盖了 Vue 3 项目开发中需要的各种各样的功能，充分体现了 Vue CLI 工具的建议式配置项目的特点。在这一步骤中同样使用上下键选择需要的功能，选中需要安装的功能后按空格键在选择和取消选择之间进行切换，选择完成后按回车键确认，然后进入下一步操作。功能选择的效果如图 10.7 所示。

在下一步操作中需要选择 Vue 的版本，选择 Vue 3，按回车键进入下一步。效果如图 10.8 所示。

在下一步操作的提示中，系统询问是否需要 class-style 的组件来支持 TypeScript。由于 Vue 3 已经对底层代码进行了重写，不需要 class 也可以很方便地进行开发，而且无须额

```
npm
Vue CLI v4.5.11
? Please pick a preset: Manually select features
? Check the features needed for your project:
 (*) Choose Vue version
 (*) Babel
>(*) TypeScript
 ( ) Progressive Web App (PWA) Support
 (*) Router
 (*) Vuex
 ( ) CSS Pre-processors
 ( ) Linter / Formatter
 ( ) Unit Testing
 ( ) E2E Testing
```

图 10.7　功能选择

```
npm
Vue CLI v4.5.11
? Please pick a preset: Manually select features
? Check the features needed for your project: Choose Vue version, Babel, TS, Router, Vu
ex
? Choose a version of Vue.js that you want to start the project with
  2.x
> 3.x (Preview)
```

图 10.8　选择 Vue 的版本

外的配置。在这一步操作中，输入英文字母 n，然后按回车键进入下一步操作。效果如图 10.9 所示。

```
npm
Vue CLI v4.5.11
? Please pick a preset: Manually select features
? Check the features needed for your project: Choose Vue version, Babel, TS, Router, Vu
ex
? Choose a version of Vue.js that you want to start the project with 3.x (Preview)
? Use class-style component syntax? (y/N) n
```

图 10.9　选择是否使用 class-style

在下一步操作中提示是否需要 Babel 和 TypeScript 结合使用，Babel 会自动添加 polyfills，并转换 JSX。因为创建的 Vue 3 项目中没有使用到 JSX，所以这一步仍然输入字母 n，按回车键进入下一步操作。效果如图 10.10 所示。

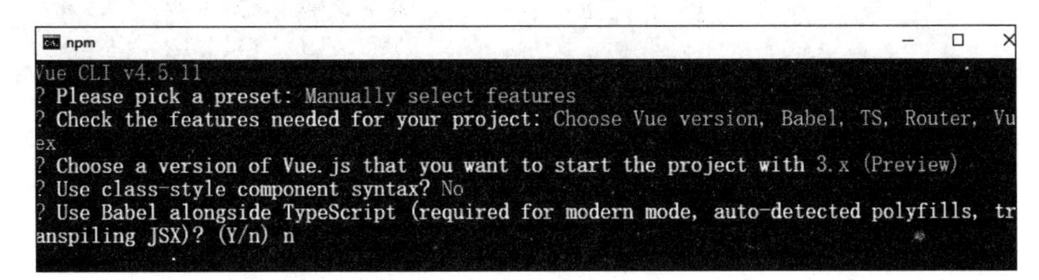

图 10.10　选择是否使用 Babel

下一步选择是否使用 history 路由模式，这里输入 y，按回车键继续下一步，效果如图 10.11 所示。

```
npm
Vue CLI v4.5.11
? Please pick a preset: Manually select features
? Check the features needed for your project: Choose Vue version, Babel, TS, Router, Vu
ex
? Choose a version of Vue.js that you want to start the project with 3.x (Preview)
? Use class-style component syntax? No
? Use Babel alongside TypeScript (required for modern mode, auto-detected polyfills, tr
anspiling JSX)? No
? Use history mode for router? (Requires proper server setup for index fallback in prod
uction) (Y/n) y
```

图 10.11 选择路由模式

下一步询问是将配置信息放到一个独立的文件中还是放到 package.json 文件中，这里选择的是放到 package.js 文件中，然后按回车键继续进入下一步。效果如图 10.12 所示。

```
npm
Vue CLI v4.5.11
? Please pick a preset: Manually select features
? Check the features needed for your project: Choose Vue version, Babel, TS, Router, Vu
ex
? Choose a version of Vue.js that you want to start the project with 3.x (Preview)
? Use class-style component syntax? No
? Use Babel alongside TypeScript (required for modern mode, auto-detected polyfills, tr
anspiling JSX)? No
? Use history mode for router? (Requires proper server setup for index fallback in prod
uction) Yes
? Where do you prefer placing config for Babel, ESLint, etc.?
  In dedicated config files
> In package.json
```

图 10.12 选择配置文件的放置位置

最后一步询问是否将前面步骤的选择保存为一个模板，方便在以后的项目创建中一键安装，这里选择 n，按回车键。效果如图 10.13 所示。

```
npm
Vue CLI v4.5.11
? Please pick a preset: Manually select features
? Check the features needed for your project: Choose Vue version, Babel, TS, Router, Vu
ex
? Choose a version of Vue.js that you want to start the project with 3.x (Preview)
? Use class-style component syntax? No
? Use Babel alongside TypeScript (required for modern mode, auto-detected polyfills, tr
anspiling JSX)? No
? Use history mode for router? (Requires proper server setup for index fallback in prod
uction) Yes
? Where do you prefer placing config for Babel, ESLint, etc.? In package.json
? Save this as a preset for future projects? (y/N) n
```

图 10.13 是否保存模板

上面的安装操作步骤完成后，会进入安装流程，这个环境需要一段时间的等待。当命令行工具中显示如图 10.14 所示的效果时，就表示项目创建成功了。

项目创建成功后，在命令行工具中输入如下命令，启动本地服务器：

```
cd vue3 - basic
npm run serve
```

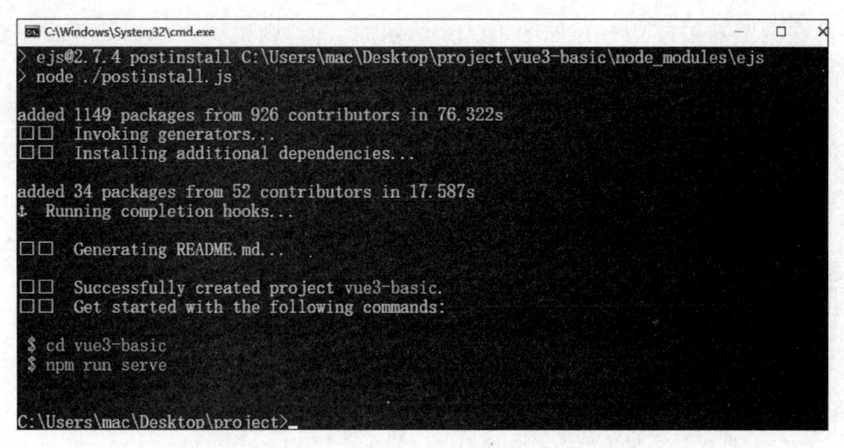

图 10.14　项目创建成功

服务器启动成功,效果如图 10.15 所示。

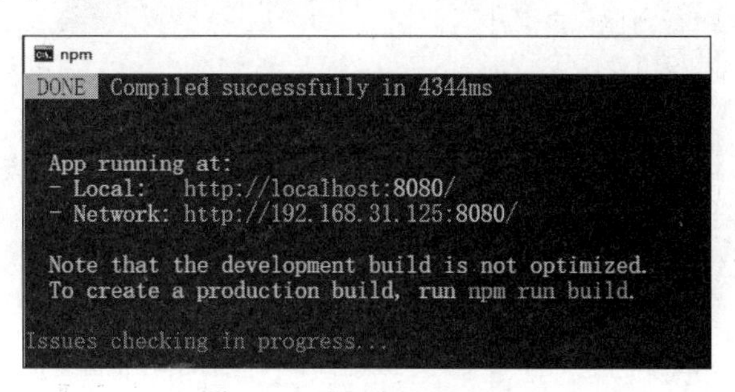

图 10.15　启动本地服务器

服务器启动成功后,在浏览器中访问 http://localhost:8080/,效果如图 10.16 所示。

图 10.16　在浏览器中访问 Vue 3 项目

如果可在浏览器中成功打开如图 10.16 所示的页面,则说明 Vue 3 的项目搭建成功了。

10.4 Vue 3 项目的目录结构

Vue 3 项目的目录结构如下。

```
- node_modules 项目的依赖管理目录
- public 公共资源管理目录
  |-- favicon.ico 站点 title 的图标
|-- index.html 站点的静态网页
- src 源码管理目录
  |-- assets 静态资源管理目录
  |-- components 公共组件管理目录
  |-- router 路由管理目录
  |-- store 状态管理目录
  |-- views 视图组件管理目录
  |-- App.vue 项目根组件
  |-- main.ts 项目入口文件
  |-- shims-vue.d.ts 定义.vue 类型的 TypeScript 配置文件
- package-lock.js 依赖管理配置文件
- package.js 依赖管理配置文件
- tsconfig.json TypeScript 配置文件
```

Vue 3 的项目目录结构与 Vue 2 的类似,唯一不同的是很多.js 文件改为了.ts 文件。其中,src 目录下的 shims-vue.d.ts 是用来定义 Vue 类型的 TypeScript 配置文件。因为.vue 结尾的 Vue 组件文件在 TypeScript 中是不能被直接识别的,所以需要使用该配置文件来讲明.vue 的类型,便于 TypeScript 进行解析。

Vue 3 中 main.ts 入口文件的源代码与 Vue 2 也有很大的差别。先来看一下 Vue 2 中的 main.js 的代码,示例代码如下:

```
import Vue from 'vue'
import App from './App.vue'
import router from './router'
import store from './store'

Vue.config.productionTip = false

new Vue({
  router,
  store,
  render: h => h(App)
}).$mount('#app')
```

在 Vue 2 中使用 new 关键字实例化 Vue 对象,然后通过构造函数将选项属性传入 Vue 实例中。而 Vue 3 对 main.js 做了修改,在 Vue 3 中使用的是.ts 类型的文件编写的入口文件,示例代码如下:

```
import { createApp } from 'vue'
import App from './App.vue'
import router from './router'
import store from './store'

createApp(App).use(store).use(router).mount('#app')
```

Vue 3 的 main. ts 文件中引入了 createApp 函数，然后引入 App. vue 组件，调用 createApp 函数来创建 Vue 实例，将所有的模块使用 Vue 实例对象进行调用，而不是像 Vue 2 中直接传入 Vue 对象的构造方法中，这是 Vue 3 做的很大改进。

10.5 Composition API 详讲

Composition API 是 Vue 的下一代主要版本中最常用的语法，它是一种全新的逻辑重用和代码组织的方法。在 Vue 2 中使用的是 Options API 的方式构建组件，如果要向 Vue 组件中添加业务逻辑，则需要先填充选项属性，例如 data、methods、computed 等。这种方式最大的缺点是，它本身并不是有效的 JavaScript 代码，需要先了解模板中可以访问哪些属性，然后使用 Vue 的编译器将这些属性转换成可以执行的代码，这样做既消耗了性能，又无法做更好的类型检查。

Composition API 设计的目的是通过将当前可用组件属性作为 JavaScript 函数暴露出来，这种机制可以基于功能的附加 API 灵活地组合组件逻辑，使 Composition API 编写的代码更易读。

下面来学习 Composition API 的语法。

10.5.1 setup()函数

setup()函数是 Vue 3 中专门新增的方法，可以理解为 Composition API 的入口。

Composition API 的主要思想是，将 Vue 组件的选项属性定义为 setup ()函数返回的 JavaScript 变量，而不是将组件的功能(例如 state、method、computed 等)定义为对象属性。

setup()函数的执行时机是在 beforeCreate 之后，created 之前。setup()函数有两个参数，第 1 个参数用于接收 props 数据。示例代码如下：

```
export default {
  props: {
    msg: {
      type: String,
      default: () => {}
    }
  },
  setup(props) {
    console.log(props);
  }
}
```

setup()函数的第 2 个参数是一个上下文对象，这个上下文对象大致包含了一些属性，示例代码如下：

```
const MyComponent = {
  setup(props, context) {
    context.attrs
    context.slots
    context.parent
    context.root
    context.emit
    context.refs
  }
}
```

这里需要注意，在 setup()函数中是无法访问 this 的。

10.5.2 reactive()函数

reactive()用来创建一个响应式对象，等价于 2.x 的 Vue.observable，示例代码如下：

```
<template>
<div>
<p @click="incment()">
            click Me!
</p>
<p>
            一:{{ state.count }} 二: {{ state.addCount }}
</p>
</div>
</template>

<script>
import { reactive } from 'vue';
export default {
    setup () {
        const state = reactive({//创建响应式数据
            count: 0,
            addCount: 0
        });

        function incment () {
            state.count++;
            state.addCount = state.count * 2;
        }

        return {
            state,
            incment
        };
```

```
        }
    };
    </script>
```

10.5.3 ref()函数

ref()函数用来给给定的值创建一个响应式的数据对象，ref()函数的返回值是一个对象，这个对象上只包含一个.value 属性。示例代码如下：

```
import { ref } from 'vue';
export default {
    setup () {
        const valueNumber = ref(0);
        const valueString = ref('hello world!');
        const valueBoolean = ref(true);
        const valueNull = ref(null);
        const valueUndefined = ref(undefined);

        return {
            valueNumber,
            valueString,
            valueBoolean,
            valueNull,
            valueUndefined
        };
    }
};
```

使用 ref()函数定义的响应式属性，在 template 中访问的方法和 Vue 2 一样，可以直接使用模板语法的形式访问，示例代码如下：

```
import { ref } from 'vue';
export default {
    setup () {
        const value = ref(1);

        return {
           value,
           msg: 'hello world!'
        };
    }
};
```

在 template 使用模板语法直接访问响应式属性，示例代码如下：

```
<template>
<p>
        {{ value }} {{ msg }}
```

```
</p>
</template>
```


下面来对比一下 Vue 3 中的 ref()函数与 Vue 2 中 data()的区别。先使用 Vue 2 的语法编写一个计算器的案例,示例代码如下:

```
//Counter.vue
export default {
  data: () => ({
    count: 0
  }),
  methods: {
    increment() {
      this.count++;
    }
  },
  computed: {
    double () {
      return this.count * 2;
    }
  }
}
```

再使用 Composition API 定义一个完全相同功能的组件,示例代码如下:

```
//Counter.vue
import { ref, computed } from "vue";

export default {
  setup() {
    const count = ref(0);
    const double = computed(() => count * 2)
    function increment() {
      count.value++;
    }
    return {
      count,
      double,
      increment
    }
  }
}
```

在上面的示例中,使用 Composition API 提供的 ref()函数定义了一个响应式变量,其作用与 Vue 2 的 data 变量几乎相同。在 Vue 3 的示例代码中,increment 方法是一个普通的 JavaScript 函数,需要更改子属性 count 的 value 才能更改响应式变量,这是因为使用 ref()函数创建的响应式变量必须是对象,以便于在传递的时候保持一致。

Composition API 提供了更方便的逻辑提取方式,还是以上面的代码为例,使用

Composition 提取 Counter. vue 组件的代码，创建 useCounter. js 文件，示例代码如下：

```
//useCounter.js
import { ref, computed } from "vue";

export default function () {
  const count = ref(0);
  const double = computed(() => count * 2)
  function increment() {
    count.value++;
  }
  return {
    count,
    double,
    increment
  }
}
```

如果要在其他组件中使用该函数，只需将模块导入组件文件并调用它，导入的模块是一个函数，该函数将从 setup()函数中返回定义的变量。示例代码如下：

```
//MyComponent.js
import useCounter from "./useCounter.js";

export default {
  setup() {
    const { count, double, increment } = useCounter();
    return {
      count,
      double,
      increment
    }
  }
}
```

这种操作还可以解决 Vue 2 中 mixins 命名冲突的问题，示例代码如下：

```
export default {
  setup () {
    const { someVar1, someMethod1 } = useCompFunction1();
    const { someVar2, someMethod2 } = useCompFunction2();
    return {
      someVar1,
      someMethod1,
      someVar2,
      someMethod2
    }
  }
}
```

Composition API 也提供了一些其他的 ref 辅助操作的函数。

1. isRef()

用来判断某个值是否为 ref 创建出来的对象，在需要展开某个值可能是 ref 创建出来的对象时使用。示例代码如下：

```
import { ref, isRef } from 'vue';
export default {
    setup () {
        const count = ref(1);
        const unwrappend = isRef(count) ? count.value : count;

        return {
           count,
           unwrappend
        };
    }
};
```

2. toRefs()

toRefs()函数可以将 reactive 创建出来的响应式对象转换为普通的对象，只不过这个对象上的每个属性节点都是 ref 类型的响应式数据。示例代码如下：

```
<template>
<p>
            <!-- 可以不通过 state.value 去获取每个属性 -->
        {{ count }} {{ value }}
</p>
</template>

<script>
import { ref, reactive, toRefs } from 'vue';
export default {
    setup () {
        const state = reactive({
            count: 0,
            value: 'hello',
        })

        return {
           ...toRefs(state)
        };
    }
};
</script>
```

3. toRef()

toRef()函数为源响应式对象上的某个属性创建一个 ref 对象，二者内部操作的是同一个数据值，更新时二者同步。与 ref 的区别是，使用 toRef()函数复制的是一份新的数据单独操作，更新时相互不影响，相当于深复制。当要将某个 prop 的 ref 传递给某个复合函数时，toRef()很有用。示例代码如下：

```
import { reactive, ref, toRef } from 'vue'

export default {
    setup () {
        const m1 = reactive({
            a: 1,
            b: 2
        })

        const m2 = toRef(m1, 'a');
        const m3 = ref(m1.a);

        const update = () => {
            //m1.a++; //m1 改变时,m2 也会改变
            //m2.value++; //m2 改变时 m1 同时改变
            m3.value++; //m3 改变的同时,m1 不会改变
        }

        return {
            m1,
            m2,
            m3,
            update
        }
    }
}
```

10.5.4 computed()计算属性

computed()函数用来创建计算属性，返回值是一个 ref 的实例。创建只读的计算属性，示例代码如下：

```
import { ref, computed } from 'vue';
export default {
    setup () {
        const count = ref(0);
        const double = computed(() => count.value + 1); //1

        double++; //Error: "double" is read - only

        return {
            count,
            double
        };
    }
};
```

在使用 computed()函数期间，传入一个包含 get()和 set()函数的对象，可以得到一个可读可写的计算属性。示例代码如下：


```
//创建一个 ref 响应式数据
const count = ref(1)

//创建一个 computed()计算属性
const plusOne = computed({
  //取值函数
  get: () => count.value + 1,
  //赋值函数
  set: val => {
    count.value = val - 1
  }
})

//为计算属性赋值的操作，会触发 set()函数
plusOne.value = 9
//触发 set()函数后，count 的值会被更新
console.log(count.value) //输出 8
```

10.5.5 Vue 3 中的响应式对象

Vue 2 中的 data 和 Vue 3 中的 ref 一样，都可以返回一个响应式对象，但是 Vue 2 中使用的是 object.defineProperty()实现响应式的，这就导致 Vue 2 的响应式出现一些限制。在 Vue 2 中新增一个响应式属性会变得很困难。

在 Vue 2 中，无法检测 property 的添加或者移除，对于已经创建的实例，Vue 是不允许动态添加根级别的响应式属性的。如果要动态添加响应式对象的属性，可以使用 Vue.set(object, propertyName, value)方法向嵌套对象中添加响应式属性，还可以使用 vm.$set 实例方法动态添加响应式属性，这也是全局 Vue.set()方法的别名。

这种操作对于一个 Vue 的初学者来讲，很多时候需要小心翼翼地去判断到底什么情况下需要用$set，什么情况下可以直接触发响应式。这给初学者带来了很多困扰。

在 Vue 3 中，这些问题都将成为过去式。Vue 3 采用了 ES6 的一个新特性，使用 Proxy 实现响应式。Proxy 对象用于定义基本操作的一个自定义行为，简单来讲，Proxy 对象就是可以让开发者对一个 JavaScript 中一切合法对象的基本操作进行自定义，然后用自定义的操作去覆盖对象的一些基本操作。

我们可以通过下面的两段代码来学习 Vue 3 中是如何使用 Proxy 进行优化的。

Vue 2 中的响应式处理，示例代码如下：

```
Object.defineProperty(data, 'count', {
    get() {},
    set() {}
})
```

Vue 3 中对于响应式的优化,示例代码如下:

```
new Proxy(data, {
    get(key) {},
    set(key, value) {}
})
```

通过上面两段代码可以看出,Proxy 是在更高维度上进行一个属性拦截修改的,先来看一下 Vue 2 的代码示例。对于给定的 data 对象,date 对象中有一个 count 属性,需要根据具体的 count 去修改 set()函数。所以,Vue 2 对于对象上的新增属性是无能为力的。

而 Vue 3 中使用 Proxy 进行拦截,这里无须知道具体的 key 是什么,拦截的是修改 data 上任意的 key 和读取 data 上任意的 key 的操作。所以,无论是已有的 key 还是新增的 key 都可以被拦截。

Proxy 更加强大之处在于,除了 getter 和 setter 对属性的拦截外,还可以拦截更多的操作符。

10.5.6 生命周期的改变

在 Vue 3 中的生命周期和在 Vue 2 中的生命周期的用法是一样的。所谓生命周期,就是一个组件从创建到销毁的全过程,会暴露出一系列的钩子函数供开发者在对应阶段进行相关的操作。

除了 Vue 2 中已有的一部分生命周期钩子,Vue 3 还增加了一些新的生命周期,可以直接导入 on××× 一族的函数来注册生命周期钩子。示例代码如下:

```
import { onMounted, onUpdated, onUnmounted } from 'vue'

const MyComponent = {
  setup() {
    onMounted(() => {
      console.log('mounted!')
    })
    onUpdated(() => {
      console.log('updated!')
    })
    onUnmounted(() => {
      console.log('unmounted!')
    })
  },
}
```

Vue 3 的生命周期钩子函数只能在 setup()期间同步使用,因为它们依赖于内部的全局状态来定位当前组件实例,不在当前组件下调用这些函数会抛出一个错误。组件实例上下文也是在生命周期钩子同步执行期间设置的,因此,在卸载组件时,在生命周期钩子内部同步创建的侦听器和计算状态也将自动删除。

Vue 3 中与 Vue 2 的生命周期相对应的组合式 API 如下。

(1) beforeCreate→使用 setup()。

(2) created→使用 setup()。

(3) beforeMount→onBeforeMount。

(4) mounted→onMounted。

(5) beforeUpdate→onBeforeUpdate。

(6) updated→onUpdated。

(7) beforeDestroy→onBeforeUnmount。

(8) destroyed→onUnmounted。

(9) errorCaptured→onErrorCaptured。

Vue 2 生命周期的初创期钩子 beforeCreate 和 created,在 Vue 3 中用 setup()替代了。除了和 2.x 生命周期等效项之外,组合式 API 还提供了以下调试钩子函数。

(1) onRenderTracked()。

(2) onRenderTriggered()。

两个钩子函数都接收一个 DebuggerEvent,与 watchEffect 参数选项中的 onTrack 和 onTrigger 类似。示例代码如下:

```
export default {
  onRenderTriggered(e) {
    debugger
    //检查哪个依赖性导致组件重新渲染
  },
}
```

10.5.7 watch()侦测变化

watch()函数用来监视某些数据项的变化,从而触发某些特定的操作,下面这个案例会实时监听 count 值的变化。示例代码如下:

```
import { ref, watch } from 'vue';
export default {
    setup () {
        const count = ref(1);

        watch(() =>{
            console.log(count.value, 'value');
        })

        setInterval(()8 =>{
            count.value++;
        },1000);
        return {
           count,
        };
    }
};
```

watch()还可以监听指定的数据源，例如监听 reactive()的数据变化。示例代码如下：

```
import { watch, reactive } from 'vue';
export default {
    setup () {
        const state = reactive({
            count: 0
        })

        watch(() => state.count, (count, prevCount) =>{
            console.log(count, prevCount); //变化后的值及变化前的值
        })

        setInterval(() =>{
            state.count++;
        },1000);

        return {
            state
        };
    }
};
```

watch()用于监听 ref 类型的数据变化，示例代码如下：

```
import { ref, watch } from 'vue';
export default {
    setup () {
        const count = ref(0);

        watch(count, (count, prevCount) =>{
            console.log(count, prevCount); //变化后的值及变化前的值
        })

        setInterval(() =>{
            count.value++;
        },1000);

        return {
            count
        };
    }
};
```

在 setup()函数内创建的 watch()监视，会在当前组件被销毁的时候自动停止。如果想要明确停止某个监视，可以调用 watch()函数的返回值。示例代码如下：

```
//创建监视，并得到停止函数
const stop = watch(() => {
```

```
  /* ... */
})

//调用停止函数,清除对应的监视
stop()
```

10.5.8 Vue 3 更好地支持 TypeScript

Vue 2 依赖于 this 上下文对象向外暴露属性,但是在设计 Vue 2 的 API 时,并没有考虑到与 TypeScript 集成。如果在 Vue 2 中想要使用 TypeScript 语法,需要使用 vue class 或者 vue extends 的方式来集成对 TypeScript 的支持。到了 Vue 3,Vue 官方团队推出了一个新的方式定义 component,这个方式称为 defineComponent。示例代码如下:

```
import { defineComponent } from 'vue';

export default defineComponent({
  setup(){
    function demo(str: String){
      console.log(str)
    }

    return {
      demo
    }
  }
});
```

10.5.9 Teleport 传送门

Teleport 是一种能够将模板移动到 DOM 中 Vue app 之外的其他位置的技术。例如,在项目中,像 modals、toast 等这样的元素,很多情况下,需要将它完全的和 Vue 应用的 DOM 剥离,这样才会更加便于项目的管理。如果类似于 modals、toast 这样的注解嵌套在 Vue 的某个组件内部,那么处理嵌套组件的定位、z-index 和样式就会变得很困难。

Teleport 很好地解决了这一类问题。下面用一个例子来讲明 Teleport 的用法。

在 index.html 中添加一个 div 元素,并指定其 id 属性值。示例代码如下:

```
<div id="app"></div>
<div id="teleport-target"></div>
```

在 HelloWorld.vue 文件中,添加 teleport 的组件代码,teleport 组件上的 to 属性要和 index.html 新增的 div 的 id 选择器保持一致。示例代码如下:

```
<button @click="showToast" class="btn">打开 toast</button>
<!-- to 属性就是目标位置 -->
```

```
<teleport to = "#teleport - target">
<div v - if = "visible" class = "toast - wrap">
<div class = "toast - msg">我是一个 Toast 文案</div>
</div>
</teleport>
```

在 HelloWorld.vue 文件中添加 script 脚本，示例代码如下：

```
import { ref } from 'vue';
export default {
  setup() {
    //toast 的封装
    const visible = ref(false);
    let timer;
    const showToast = () => {
      visible.value = true;
      clearTimeout(timer);
      timer = setTimeout(() => {
        visible.value = false;
      }, 2000);
    }
    return {
      visible,
      showToast
    }
  }
}
```

在上面的示例中，使用 teleport 组件，通过 to 属性指定该组件渲染的位置与< div id='app'>同级，但是 teleport 的状态 visible 又是完全由内部 Vue 组件控制的。

10.5.10 Suspense 异步请求

Vue 3 新增了 Suspense 组件，可以允许应用程序在等待异步组件时渲染一些后备内容，帮助开发者创建一个平滑的用户体验。Suspense 组件非常容易理解，也不需要任何额外的导入。

例如，有一个异步的 ArticleInfo.vue 的组件，其中 setup()方法是异步的，用于返回加载用户的数据。示例代码如下：

```
async function getArticleInfo() {
  //调用一些异步 API
  return { article }
}export default {
  async setup () {
    var { article } = await getArticleInfo()
    return {
      article
```

```
    }
}}
```


再创建一个 ArticlePost.vue 组件,在该组件中包含 ArticleInfo.vue 组件。如果要在等待组件获取数据并解析时显示正在加载的内容,只需 3 步就可以实现 Suspense。

(1) 将异步组件包装在< template #default >标记中。

(2) 在 Async 组件的旁边添加一个兄弟姐妹,标签为< template #fallback >。

(3) 将两个组件都包装在< suspense >组件中。

使用插槽,Suspense 将渲染后备内容,直到默认内容准备就绪。然后,它将自动切换以显示异步组件。示例代码如下:

```
<Suspense>
<template #default>
<article-info/>
</template>
<template #fallback>
<div>正在拼了命的加载…</div>
</template>
</Suspense>
```

10.5.11 全局 API 修改

Vue 2 有许多全局 API 和配置,这些 API 和配置可以全局改变 Vue 的行为。例如,要创建全局组件,可以使用 Vue.component 这样的 API。示例代码如下:

```
Vue.component('button-counter', {
  data: () => ({
    count: 0
  }),
  template: '<button @click="count++">Clicked {{ count }} times.</button>'
})
```

虽然这种声明方式很方便,但它也会导致一些问题。在 Vue 2 中只能通过 new Vue 创建根 Vue 实例,从同一个 Vue 构造函数创建的每个根实例共享相同的全局配置,因此,在测试期间,全局配置很容易意外地污染其他测试用例。

Vue 3 中增加了 createApp 这个新的全局 API,调用 createApp 返回一个应用实例。示例代码如下:

```
import { createApp } from 'vue'
const app = createApp({})
```

任何全局改变 Vue 行为的 API 在使用了 createApp 之后,都会转移到应用实例上。如表 10.1 所示,Vue 2 中的全局 API 转移到了 Vue 3 中的实例 API 上。

表 10.1　Vue 2 全局 API 与 Vue 3 实例 API 对照表

Vue 2 全局 API	Vue 3 实例 API(app)
Vue. config	app. config
Vue. config. productionTip	removed
Vue. config. ignoredElements	app. config. isCustomElement
Vue. component	app. component
Vue. directive	app. directive
Vue. mixin	app. mixin
Vue. use	app. use
Vue. prototype	app. config. globalProperties

所有其他不全局改变行为的全局 API 现在被命名为 exports。

项目实战篇

第 11 章　实战——Vue 2 仿“京东商城”App

本章将介绍一款仿“京东商城”商品信息展示的电商类 App。该案例基于 Vue 2、Vue Router、Webpack、ES6 等技术栈实现的一款 App，很适合初学者进行学习。

11.1　项目概述

此项目是一款仿“京东商城”的商品信息展示 App，主要实现了以下功能。

(1) 商城首页轮播效果，热销商品展示，公共底部导航。

(2) 搜索页面搜索关键词智能提示，保存搜索记录。

(3) 商品一级分类与二级分类导航的展示。

(4) 商品列表页面的商品展示。

(5) 单击商品添加购物车。

(6) 购物车页面的商品购买数量加减，统计购买商品的总价。

11.1.1　开发环境

首先需要安装 Node.js 12 及以上版本，因为 Node.js 中已经继承了 NPM，所以无须单独安装 NPM，然后安装 Vue 脚手架(Vue-CLI)及创建项目。

项目的调试使用浏览器的控制台进行，在浏览器中按下 F12 键，然后单击“切换设备工具栏”，进入移动端的调试界面，可以选择相应的设备进行调试，效果如图 11.1 所示。

11.1.2　项目结构

项目结构如图 11.2 所示，其中 src 文件夹是项目的源文件目录，src 文件夹下的项目结构如图 11.3 所示。

项目结构中主要文件说明如下。

(1) dist：项目打包后的静态文件存放目录。

(2) node_modules：项目依赖管理目录。

(3) public：项目的静态文件存放目录，也是本地服务器的根目录。

(4) src：项目源文件存放目录。

(5) package.json：项目 NPM 配置文件。

src 文件夹目录说明如下。

(1) assets：静态资源文件存放目录。

(2) components：公共组件存放目录。

图 11.1　项目调试效果

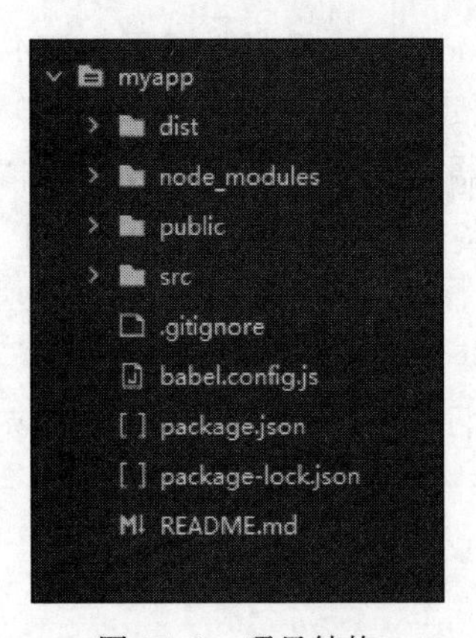

图 11.2　项目结构

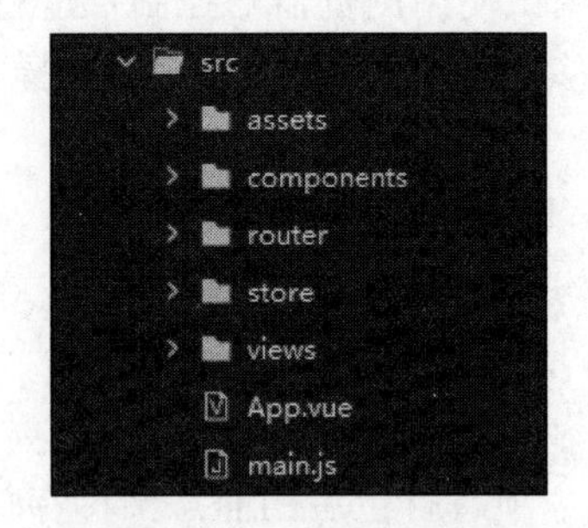

图 11.3　src 文件夹

(3) router：路由配置文件存放目录。

(4) store：状态管理配置存放目录。

(5) views：视图组件存放目录。

(6) App.vue：项目的根组件。

(7) main.js：项目的入口文件。

项目的所有页面组件存放在 src/views 目录下，该目录下的文件说明如下。

（1）Carts. vue：定义购物车页面的视图组件。

（2）Classify. vue：定义商品分类页面的视图组件。

（3）GoodsList. vue：定义商品列表页面的视图组件。

（4）Index. vue：定义项目的公共底部导航与三级路由的视图组件。

（5）Main. vue：定义商城的首页布局的视图组件。

（6）Myself. vue：定义“我的”页面的视图组件。

（7）Search. vue：定义搜索页面的视图组件。

11.2 入口文件

项目的入口文件有 index. html、main. js 和 App. vue 3 个，这些入口文件的具体内容介绍如下。

11.2.1 项目入口页面

index. html 是项目默认的主渲染页面文件，主要用于 Vue 实例挂载点的声明与 DOM 渲染，代码如下：

```
<!DOCTYPE html>
<html lang="en">
  <head>
    <meta charset="UTF-8">
    <meta http-equiv="X-UA-Compatible" content="IE=edge">
    <meta name="viewport" content="width=device-width,initial-scale=1.0">
    <link rel="icon" href="<%= BASE_URL %>favicon.ico">
    <title><%= htmlWebpackPlugin.options.title %></title>
  </head>
  <body>
    <div id="app"></div>
  </body>
</html>
```

11.2.2 程序入口文件

main. js 是程序的入口文件，主要用于加载各种公共组件和初始化 Vue 实例。本项目中的路由设置和引用的 Vant UI 组件库就是在该文件中定义的，代码如下：

```
import Vue from 'vue'
import App from './App.vue'
import router from './router'
import store from './store'
import Vant from 'vant'
import 'vant/lib/index.css'
import axios from 'axios'
```

```
//引入 Vant 组件库
Vue.use(Vant);
Vue.config.productionTip = false
//引入 axios 模块
Vue.prototype.$axios = axios

new Vue({
  router,
  store,
  render: h => h(App)
}).$mount('#app')
```

11.2.3 组件入口文件

App.vue 是项目的根组件，所有的页面都是在 App.vue 下面进行切换的，所有的页面组件都是 App.vue 的子组件。在 App.vue 组件内只需使用 <router-view> 组件作为占位符，就可以实现各个页面的引入，代码如下：

```
<template>
  <div>
    <router-view></router-view>
  </div>
</template>
```

11.3 项目组件

项目中所有页面组件都在 views 文件夹中定义，具体组件内容如下。

11.3.1 底部导航组件

在仿"京东商城"App 中，底部导航使用的是 Vant UI 组件库中的 Tabbar 标签栏组件，组件引入代码如下：

```
import Vue from 'vue';
import { Tabbar, TabbarItem } from 'vant';

Vue.use(Tabbar);
Vue.use(TabbarItem);
```

公共底部导航组件的代码如下：

```
<van-tabbar v-model="active" route active-color="#F6230E"
inactive-color="#8B8B8B">
        <van-tabbar-item icon="home-o"
                        to='/main'>
                        首页
```

```
        </van-tabbar-item>
        <van-tabbar-item icon="apps-o"
                     to='/classify'>
                     分类
        </van-tabbar-item>
        <van-tabbar-item icon="shopping-cart-o"
                     to='/carts'
                     :badge="cartList.length">
                     购物车
        </van-tabbar-item>
        <van-tabbar-item icon="contact"
                     to='/myself'>
                     我的
        </van-tabbar-item>
    </van-tabbar>
```

运行效果如图 11.4 所示。

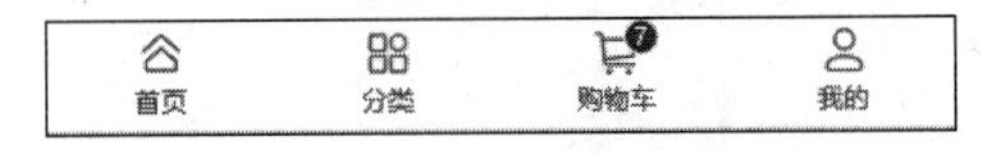

图 11.4　底部导航效果图

11.3.2　商城首页

首页是项目的核心页面之一，在首页中主要展示商城的重要商品信息，首页的布局也是相对比较复杂的。仿"京东商城"App 首页效果如图 11.5 所示。

图 11.5　仿"京东商城"App 首页效果

在整个项目中，一级路由编写在 App.vue 根组件中，关于首页的二级路由则编写在 Index.js 组件中。

Index.vue 文件代码如下：


```
<template>
  <div>
      <router-view></router-view>
      <van-tabbar v-model="active"
                  route
                  active-color="#F6230E"
                  inactive-color="#8B8B8B">
        <van-tabbar-item icon="home-o"
                    to='/main'>
                    首页
            </van-tabbar-item>
        <van-tabbar-item icon="apps-o"
                    to='/classify'>
                    分类
            </van-tabbar-item>
        <van-tabbar-item icon="shopping-cart-o"
                    to='/carts'
                    :badge="cartList.length">
                    购物车
            </van-tabbar-item>
        <van-tabbar-item icon="contact"
                    to='/myself'>
                    我的
            </van-tabbar-item>
    </van-tabbar>
  </div>
</template>

<script>
export default {
    data(){
        return {
            active: 0,
            cartList: []
        }
    },
    created(){
        //获取购物车数据
        let list = localStorage.cartList
        if(list){
        this.cartList = JSON.parse(list)
        }
    }
}
</script>
```

在上面的代码中，单击底部导航的“首页”按钮，会将对应的首页视图组件 Main. vue 文件加载到< router-view >组件中。

Main. vue 文件代码如下：

```
<template>
  <div>
    <van-search placeholder="请输入搜索关键词"
                shape="round"
                background="#E43130"
                show-action
                @focus="searchFocus"
                style="position: fixed;width:100%;top: 0;z-index: 999;"
        >
        <template #left>
            <van-icon name="wap-nav"
                    color="#fff"
                    size="25px"
                    style="margin-right: 8px;" />
        </template>
        <template #label>
            <span class="search-logo">JD</span>
        </template>
        <template #action>
            <span style="font-size: 16px;color:#fff;">登录</span>
        </template>
      </van-search>

      <!-- 轮播广告 -->
      <div class="swiper-banner" style="margin-top: 70px">
        <van-swipe class="my-swipe"
                :autoplay="3000"
                indicator-color="white">
          <van-swipe-item v-for="URL in imgs" :key="URL">
            <img :src="URL" width="100%" >
          </van-swipe-item>
        </van-swipe>
      </div>

      <!-- 宫格 -->
      <div class="swiper-banner">
        <van-swipe class="my-swipe"
                :autoplay="3000"
                indicator-color="red"
                :loop="false"
                style="padding: 10px 0px" >
          <van-swipe-item>
            <van-grid :column-num="5" :border="false">
              <van-grid-item v-for="(item,index) in navData"
                            :key="index"
```

```
                        :text = "item.title"
                        v - show = "index < 10">
                <template #icon>
                    <img :src = "item.imgURL" width = "50px">
                </template>
              </van - grid - item>
            </van - grid>
          </van - swipe - item>
          <van - swipe - item>
            <van - grid :column - num = "5" :border = "false">
              <van - grid - item v - for = "(item,index) in navData"
                        :key = "index"
                        :text = "item.title"
                        v - show = "index >= 10 && index < 20">
                <template #icon>
                    <img :src = "item.imgURL" width = "50px">
                </template>
              </van - grid - item>
            </van - grid>
          </van - swipe - item>
        </van - swipe>
      </div>

      <!-- 商品推荐 -->
      <img src = "/imgs/tuijian.png" width = "100%">

      <div class = "goods - list">
        <goods - card v - for = "i in 10"
                :key = "i"
                img = "/imgs/goods.jpg"
                title = "商品标题"
                :price = "10">
        </goods - card>
      </div>

  </div>
</template>

<script>
import GoodsCard from '@/components/GoodsCard'
import imgURLs from '@/assets/imgs.js'
export default {
  components: {
    'goods - card': GoodsCard
  },
  data(){
    return {
      imgs: [],
      navData: []
    }
```

```
  },
  created() {
      this. $ axios.get( '/data/home.json' ).then(res =>{
              this.imgs = res.data.bannerImgs;
              this.navData = res.data.icons;
      }).catch(err =>{
              console.error(err)
      })
  },
  methods: {
    searchFocus(){
            this. $ router.push({
                path: '/search'
            })
        }
  }
}
</script>

<style scoped>
.search-logo{
    font-size: 18px;
    color: #EA4546;
    font-weight: bold;
    padding: 0px 10px;
    border-right: 1px solid #E7E7E7;
}
.swiper-banner{
  width: 90%;
  margin: 10px auto;
  border-radius: 10px;
  overflow: hidden;
}
.goods-list{
  display: flex;
  justify-content: space-between;
  box-sizing: border-box;
  padding: 0px 10px;
  flex-wrap: wrap;
  margin-bottom: 80px;
}
</style>
```

在上面的代码中，轮播图和宫格使用了本地 JSON 数据，通过 axios 获取本地数据，本地 JSON 文件代码存放在 home.json 文件中。

/public/data/home.json 文件代码如下：

```
{
    "bannerImgs": [
            "/imgs/banner01.jpg",
            "/imgs/banner02.jpg",
            "/imgs/banner03.jpg"
    ],
    "icons": [
            {
                    "title": "京东超市",
                    "imgURL": "/imgs/icon-01.png"
            },
            {
                    "title": "数码电器",
                    "imgURL": "/imgs/icon-02.png"
            },
            {
                    "title": "京东服饰",
                    "imgURL": "/imgs/icon-03.png"
            },
            {
                    "title": "京东生鲜",
                    "imgURL": "/imgs/icon-04.png"
            },
            {
                    "title": "京东到家",
                    "imgURL": "/imgs/icon-05.png"
            },
            {
                    "title": "充值缴费",
                    "imgURL": "/imgs/icon-06.png"
            },
            {
                    "title": "9.9 元拼",
                    "imgURL": "/imgs/icon-07.png"
            },
            {
                    "title": "领券",
                    "imgURL": "/imgs/icon-08.png"
            },
            {
                    "title": "领金贴",
                    "imgURL": "/imgs/icon-09.png"
            },
            {
                    "title": "PLUS 会员",
                    "imgURL": "/imgs/icon-10.png"
            },
            {
```

```
                "title": "京东国际",
                "imgURL": "/imgs/icon-11.png"
            },
            {
                "title": "京东拍卖",
                "imgURL": "/imgs/icon-12.png"
            },
            {
                "title": "唯品会",
                "imgURL": "/imgs/icon-13.png"
            },
            {
                "title": "玩 3C",
                "imgURL": "/imgs/icon-14.png"
            },
            {
                "title": "沃尔玛",
                "imgURL": "/imgs/icon-15.png"
            },
            {
                "title": "美妆馆",
                "imgURL": "/imgs/icon-16.png"
            },
            {
                "title": "京东旅行",
                "imgURL": "/imgs/icon-17.png"
            },
            {
                "title": "拍拍二手",
                "imgURL": "/imgs/icon-18.png"
            },
            {
                "title": "物流查询",
                "imgURL": "/imgs/icon-19.png"
            },
            {
                "title": "全部",
                "imgURL": "/imgs/icon-20.png"
            }
        ]
    }
```

11.3.3 搜索页面

搜索页面的功能稍微复杂一些，当在搜索框输入关键词时，需要智能提示，然后单击"搜索"按钮，跳转到搜索结果页面。对于搜索的所有关键词，要保存到本地存储文件中，在搜索页面以"搜索记录"的形式进行展示，并实现对搜索记录进行删除的功能。搜索页面显示效果如图 11.6 所示，搜索智能提示效果如图 11.7 所示。

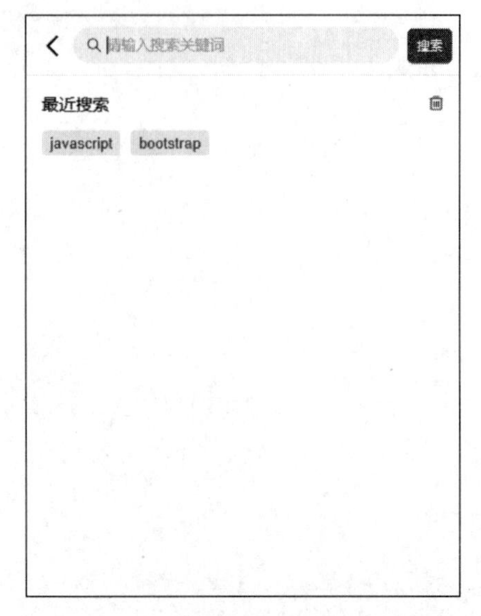

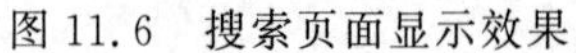
图 11.6　搜索页面显示效果

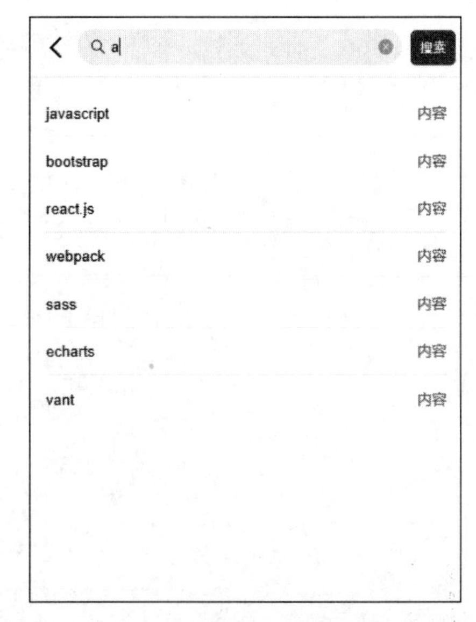

图 11.7　搜索智能提示效果

搜索框使用 Vant UI 的 Search 搜索组件，在使用组件前要引入 Search 组件，代码如下：

```
import Vue from 'vue';
import { Search } from 'vant';

Vue.use(Search);
```

Search.vue 组件代码如下：

```
<template>
  <div>
      <van-search v-model="searchValue"
                  placeholder="请输入搜索关键词"
                  shape="round"
                  background="#fff"
                  show-action
                  autofocus
                  @search="onSearch"
            >
            <template #left>
                <van-icon @click="pageBack"
                          name="arrow-left"
                          size="25px"
                          style="margin-right: 8px;" />
            </template>
            <template #action>
                <van-button color="#E93B3D"
                            size="small"
                            style="border-radius: 5px"
                            @click="onSearch">
                            搜索
```

```
                </van-button>
            </template>
        </van-search>

        <!-- 搜索记录 -->
        <div class="search-history">
            <div class="search-history-title">
                <span>最近搜索</span>
                <van-icon name="delete" @click="clearHistory" />
            </div>
            <div class="search-history-list">
                <van-tag v-for="(item,index) in historyList"
                          :key="index"
                        color="#F0F2F5"
                        text-color="#68687F"
                        size="large"
                        style="margin:0px 10px 10px 0px;"
                        >
                        {{item}}
                </van-tag>
            </div>
        </div>

        <!-- 搜索智能提示区域 -->
        <div class="kw-list" v-show="showKwList">
            <van-cell v-for="kw in showList"
                        :key="kw"
                        :title="kw"
                        value="内容"
                        @click="onSearch(kw)" />
        </div>
    </div>
</template>

<script>
export default {
    data(){
        return {
            searchValue: '',         //搜索的内容
            showKwList: false ,      //控制智能搜索区域的显示
            historyList: [] ,        //搜索记录列表
            showList: [],            //要展示的内容
            data: [
                "html",
                "css",
                "JavaScript",
                "jquery",
                "node.js",
                "Vue.js",
                "swiper",
```

```
                "bootstrap",
                "php",
                "MongoDB",
                "MySQL",
                "react.js",
                "GitHub",
                "glup",
                "Webpack",
                "sass",
                "echarts",
                "vant"
            ]
        }
    },
    created(){ //初始化搜索记录
        let historyList = localStorage.historyList
        if(historyList){
            this.historyList = JSON.parse(historyList)
        }
    },
    watch: {
        searchValue(kw){
            this.showKwList = kw.length > 0 ? true : false
            this.showList = this.data.filter(item =>{
                return item.includes(kw)
            })
        }
    },
    methods: {
        pageBack(){
            //返回上一页
            window.history.back()
        },
        onSearch(kw){                                   //搜索事件
            let keyword = ''
            if(typeof kw === 'string'){                 //单击智能提示或回车搜索
                keyword = kw
            }else if(typeof kw === 'object') {          //单击搜索按钮
                if(this.searchValue.trim() == '') return
                keyword = this.searchValue
            }

            //执行搜索功能
            this.$router.push({
                path: '/list',
                query: {
                    kw: keyword
                }
            })
            //保存搜索记录
```

```
            this.saveSeachKw(keyword)
        },
        saveSeachKw(kw){                    //保存搜索关键字
            if(this.historyList.includes(kw)){
                let index = this.historyList.indexOf(kw)
                this.historyList.splice(index,1)
                this.historyList.unshift(kw)
            }else{
                this.historyList.unshift(kw)
            }
            //将更新后的 historyList 同步到本地
            localStorage.historyList = JSON.stringify(this.historyList)
        },
        clearHistory(){                     //清空搜索记录
            this.$dialog.confirm({
                message: '确定要清空吗?',
                closeOnClickOverlay: true,
                confirmButtonText: '清空'
            }).then(() =>{
                this.historyList = []
                localStorage.historyList = JSON.stringify(this.historyList)
            }).catch(() =>{
            })
        }
    }
}
</script>

<style scoped>
/* 搜索记录区域 */
.search-history{
    border-top: 1px solid #F2F2F2;
}
.search-history-title{
    height: 50px;
    display: flex;
    align-items: center;
    justify-content: space-between;
    box-sizing: border-box;
    padding: 0px 15px;
}
.search-history-list{
    box-sizing: border-box;
    padding: 0px 15px;
}
/* 智能搜索提示区域 */
.kw-list{
    background: #fff;
    position: absolute;
    width: 100%;
```

```
    top: 67px;
    min-height: 150px;
}
</style>
```


11.3.4 分类导航页面

商品分类页面的头部和底部都使用公共的 UI 组件，底部使用的是 Vant UI 组件库中的 Tabbar 标签栏组件，头部则使用 Search 搜索组件。在分类页面，主要实现一级商品分类导航和二级商品分类导航的布局与关联。分类导航页面显示效果如图 11.8 所示。

图 11.8　分类导航页面显示效果

在分类页面中，所有的分类数据都通过本地模拟出来，然后使用 axios 模块异步加载本地服务器上的 JSON 数据。

Classify.vue 分类页面核心部分的代码如下：

```
<template>
  <div class="nav-body">
    <!-- 一级分类导航 -->
      <ul class="navOne">
        <li v-for="(item) in navs"
                      :key="item.cid"
                      @click="clickNavOne(item)">
          <img :src="item.cpic" width="30px" height="30px">
          {{item.cname}}
```

```
        </li>
      </ul>
<!-- 二级分类导航 -->
      <ul class="sonNav">
        <li v-for="(item) in sonNav"
                    :key="item.subcid"
                    @click="toPage">
          <img :src="item.scpic" width="50px" height="50px">
          {{item.subcname}}
        </li>
      </ul>
  </div>
</template>

<script>
export default {
  data(){
    return {
      navs: [],                //所有的导航数据
      sonNav: []               //二级导航数据
    }
  },
  created(){
    //获取服务器端数据
    this.$axios.get("http://localhost:8080/data/navs.json")
    .then((res) =>{
      console.log(res.data.data.data)
      this.navs = res.data.data.data
      this.sonNav = this.navs[0].subcategories
    })
  },
  methods: {
    clickNavOne(item){
      this.sonNav = item.subcategories
    },
    toPage(){                  //跳转
      this.$router.push({
        path: '/list'
      })
    }
  }
}
</script>

<style scoped>
/* 垂直导航样式 */
.nav-body{
  display: flex;
}
.navOne li{
```

```
  margin: 10px 0px;
  text-indent: 10px;
  display: flex;
  align-items: center;
  height: 40px;
}
.sonNav li{
  margin: 10px 0px;
  display: flex;
  flex-direction: column;
  justify-content: center;
  align-items: center;
  width: 50%;
}
.navOne{
  position: relative;
  flex: 1.5;
  border-right: 1px solid #eee;
  height: 600px;
  overflow-y: scroll;
}
/*隐藏滚动条 */
::-webkit-scrollbar{
  display: none;
}
.sonNav{
  flex: 3;
  display: flex;
  flex-wrap: wrap;
}
</style>
```

11.3.5 商品列表页面

在单击商品分类和关键词搜索后，会跳转到商品列表页面，该页面的所有数据都通过条件查询获取，例如，关键词查询、分类查询等。商品列表页面显示效果如图 11.9 所示。

在商品列表页面中，头部使用了 Vant UI 组件库中的 NavBar 导航栏组件，该 UI 组件的引入代码如下：

```
import Vue from 'vue';
import { NavBar } from 'vant';

Vue.use(NavBar);
```

图 11.9 商品列表页面显示效果

GoodsList.vue 商品列表页面的代码如下：

```
<template>
  <div>
    <van-nav-bar
      title="商品列表"
      left-text="返回"
      :right-text="购物车(${cartList.length})"
      left-arrow
      fixed
      @click-left="onClickLeft"
      @click-right="onClickRight"
    />
    <div class="list-body">
        <goods-card v-for="(item) in goods"
                    :key="item.goodsId"
                    :title="item.title"
                    :price="item.originalPrice"
                    :img="item.mainPic"
                    @click="addCarts(item)"
              >
        </goods-card>
    </div>
  </div>
</template>

<script>
import GoodsCard from '@/components/GoodsCard'
```

```
export default {
  components: {
    'goods - card': GoodsCard
  },
  data(){
    return {
      goods: [],                //所有商品数据
      cartList: []              //购物车商品数据
    }
  },
  watch: {
    cartList: {
      handler(list){
        //保存到本地
        localStorage.cartList = JSON.stringify(list)
      },
      deep: true
    }
  },
  created(){
    //获取商品数据
    this.$axios.get('/data/list.json').then(res =>{
      //console.log(res.data.data.data.list)
      this.goods = res.data.data.data.list
    })

    //获取购物车数据
    let list = localStorage.cartList
    if(list){
      this.cartList = JSON.parse(list)
    }
  },
  methods: {
    onClickLeft(){              //返回按钮
      window.history.back()
    },
    onClickRight(){             //查看购物车按钮
      this.$router.push({
        path: '/carts'
      })
    },
    addCarts(item){             //添加购物车

      //判断是否重复添加,true 为重复添加,false 为第一次添加
      let double = false
      this.cartList.map(cart =>{
        if(cart.goods.goodsId == item.goodsId){
          //该商品已经添加过购物车了
          //把该商品的购买数量 +1
          cart.num ++
```

```
          double = true
          return
        }
      })

      //第一次添加
      if(!double){
        this.cartList.push({
          goods: item,
          num: 1
        })
      }
    }
  }
}
</script>

<style scoped>
.list-body{
  display: flex;
  justify-content: space-between;
  flex-wrap: wrap;
  margin-top: 50px;
}
</style>
```

11.3.6 购物车页面

单击商品卡片，可以将该商品加入购物车，在本项目案例中，购物车使用了 localStorage 本地存储实现对购物车商品的保存。购物车的头部使用了 Vant UI 组件库中的 NavBar 导航栏组件，引入代码如下：

```
import Vue from 'vue';
import { NavBar } from 'vant';
Vue.use(NavBar);
```

在购物车页面中，商品列表的每个商品卡片使用了 Vant UI 组件库中的 Card 卡片业务组件，引入代码如下：

```
import Vue from 'vue';
import { Card } from 'vant';

Vue.use(Card);
```

购物车页面底部的购买商品总价计算与全选按钮使用了 Vant UI 组件库中的 SubmitBar 提交订单栏业务组件，引入代码如下：

```
import Vue from 'vue';
import { SubmitBar } from 'vant';

Vue.use(SubmitBar);
```

购物车页面的商品展示效果如图 11.10 所示，单击全选按钮后，选中所有商品并计算商品总价，效果如图 11.11 所示。

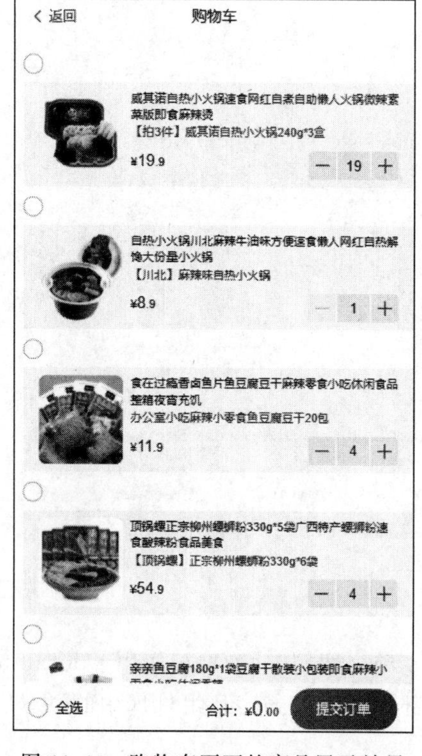

图 11.10　购物车页面的商品展示效果

图 11.11　商品全选效果

Carts.vue 购物车页面的核心代码如下：

```
<template>
  <div>
      <van-nav-bar
          title="购物车"
          fixed
          left-text="返回"
          left-arrow
          @click-left="onClickLeft"
        />
      <!-- 购物车商品列表 -->
      <div style="margin-top: 60px;margin-bottom: 70px">
        <!-- 复选框组 -->
        <van-checkbox-group v-model="result">
```

```
        <div class = "goods - item"
                          v - for = "item in cartList"
                          :key = "item.goodsId">
          <van - swipe - cell >
            <van - checkbox :name = "item"></van - checkbox >
            <van - card
                :price = "item.goods.originalPrice"
                :desc = "item.goods.dtitle"
                :title = "item.goods.title"
                :thumb = "item.goods.mainPic"
              >
                <template #num >
                  <van - stepper v - model = "item.num" />
                </template >
              </van - card >
            <template #right >
              <van - button square
                  text = "删除"
                  type = "danger"
    class = "delete - button" />
            </template >
          </van - swipe - cell >
        </div >
      </van - checkbox - group >
    </div >
    <!-- 提交订单 -->
    <van - submit - bar :price = "totalPrice" button - text = "提交订单">
      <van - checkbox v - model = "selAll" @click = "handleSellAll">
                          //全选
            </van - checkbox >
    </van - submit - bar >
  </div >
</template >

<script >
export default {
  data(){
    return {
      cartList: [] ,          //购物车数据
      selAll: false ,         //全选状态
      totalPrice: 0 ,         //总价,单位分
      result: [] ,            //当前被选中的商品对象数组
    }
  },
  watch: {
    cartList: {
      handler(list){
                          //保存到本地
        localStorage.cartList = JSON.stringify(list)
      },
```

```
        deep: true
      },
      result: { //监听商品选中
        handler(list){
          //判断当前选中的商品数量是否和购物车中一致
          if(list.length == this.cartList.length){
            this.selAll = true
          }else{
            this.selAll = false
          }
          this.comTotalPrice()
        },
        deep:true
      }
    },
    created(){
      //获取购物车数据
      let list = localStorage.cartList
      if(list){
        this.cartList = JSON.parse(list)
      }
    },
    methods: {
      onClickLeft(){ //返回
        window.history.back()
      },
      comTotalPrice(){ //计算总价
          let totalPrice = 0
          this.result.map(item =>{
            totalPrice += item.goods.originalPrice * item.num
          })

          this.totalPrice = totalPrice * 100
      },
      handleSellAll(){ //全选按钮
        if(this.selAll){
          this.result = this.cartList
        }else{
          this.result = []
        }
      }
    }
  }
</script>

<style scoped>
.goods-item{
  margin: 10px 0px;
  box-sizing: border-box;
  padding-left: 10px;
```

```
  display: flex;
}
.delete-button{
  height: 100%;
}
</style>
```

第12章 实战——Vue 2 仿“饿了么”App

本章将介绍一款仿“饿了么”商家页面的 App。该案例基于 Vue 2、Vue Router、Webpack、ES6 等技术栈实现的一款外卖类 App，适合初学者进行学习。

12.1 项目概述

该项目是一款仿“饿了么”商家页面的外卖类 App，主要有以下功能。

（1）商品导航。

（2）商品列表使用手势上下滑动。

（3）购物车中商品的添加和删除操作。

（4）单击商品查看详情。

（5）商家评论。

（6）商家信息。

12.1.1 开发环境

首先需要安装 Node.js 12 及以上版本，因为 Node.js 中已经继承了 NPM，所以无须单独安装 NPM，然后安装 Vue 脚手架（Vue-CLI）及创建项目。

项目的调试使用浏览器的控制台进行，在浏览器中按下 F12 键，然后单击“切换设备工具栏”，进入移动端的调试界面，可以选择相应的设备进行调试。项目效果如图 12.1 所示。

12.1.2 项目结构

项目结构如图 12.2 所示，其中 src 文件夹是项目的源文件目录，src 文件夹下的项目结构如图 12.3 所示。

项目结构中主要文件说明如下。

（1）dist：项目打包后的静态文件存放目录。

（2）node_modules：项目依赖管理目录。

（3）public：项目的静态文件存放目录，也是本地服务器的根目录。

（4）src：项目源文件存放目录。

（5）package.json：项目 NPM 配置文件。

src 文件夹目录说明如下。

（1）assets：静态资源文件存放目录。

（2）components：公共组件存放目录。

图 12.1　项目效果图

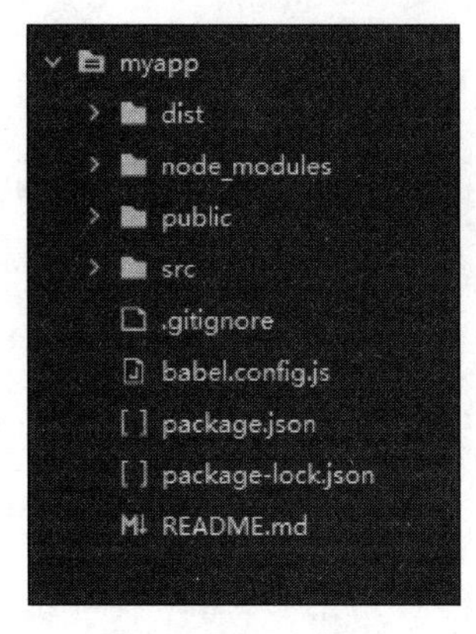

图 12.2　项目结构

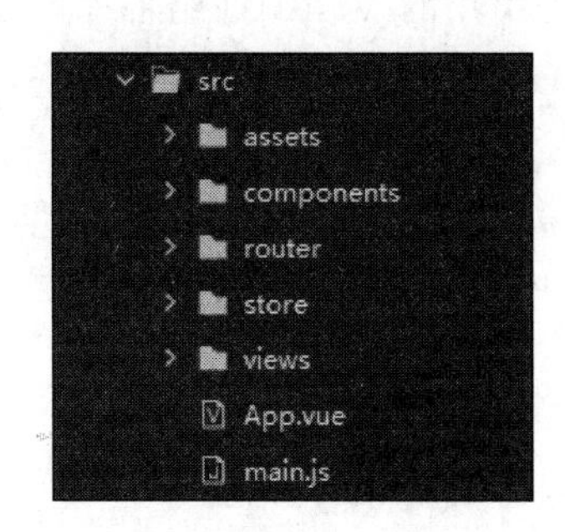

图 12.3　src 文件夹

（3）router：路由配置文件存放目录。

（4）store：状态管理配置存放目录。

（5）views：视图组件存放目录。

（6）App. vue：项目的根组件。

（7）main. js：项目的入口文件。

12.2 入口文件

项目的入口文件有 index.html、main.js 和 App.vue 3 个，这些入口文件的具体内容如下。

12.2.1 项目入口页面

index.html 是项目默认的主渲染页面文件，主要用于 Vue 实例挂载点的声明与 DOM 渲染，代码如下：

```
<!DOCTYPE html>
<html lang="en">
  <head>
    <meta charset="UTF-8">
    <meta http-equiv="X-UA-Compatible" content="IE=edge">
    <meta name="viewport" content="width=device-width, initial-scale=1.0">
    <link rel="icon" href="<%= BASE_URL %>favicon.ico">
    <title><%= htmlWebpackPlugin.options.title %></title>
  </head>
  <body>
    <div id="app"></div>
  </body>
</html>
```

12.2.2 程序入口文件

main.js 是程序的入口文件，主要用于加载各种公共组件和初始化 Vue 实例。本项目中的路由设置和引用的 Vant UI 组件库就是在该文件中定义的，代码如下：

```
import Vue from 'vue'
import App from './App.vue'
import './cube-ui'
import './register'

import 'common/stylus/index.styl'

Vue.config.productionTip = false

new Vue({
  render: h => h(App)
}).$mount('#app')
```

本项目案例使用了 Cube UI 组件库，在项目 src 目录下创建 cube-ui.js 文件，用于引入项目中要用到的组件，代码如下：

```
import Vue from 'vue'
import {
  Style,
  TabBar,
  Popup,
  Dialog,
  Scroll,
  Slide,
  ScrollNav,
  ScrollNavBar
} from 'cube-ui'

Vue.use(TabBar)
Vue.use(Popup)
Vue.use(Dialog)
Vue.use(Scroll)
Vue.use(Slide)
Vue.use(ScrollNav)
Vue.use(ScrollNavBar)
```

12.2.3 组件入口文件

App.vue 是项目的根组件，所有的页面都在 App.vue 下面进行切换，所有的页面组件都是 App.vue 的子组件。在 App.vue 组件内只需使用 <router-view> 组件作为占位符，就可以实现各个页面的引入，代码如下：

```
<template>
  <div id="app" @touchmove.prevent>
    <v-header :seller="seller"></v-header>
    <div class="tab-wrapper">
      <tab :tabs="tabs"></tab>
    </div>
  </div>
</template>

<script>
  import qs from 'query-string'
  import { getSeller } from 'api'
  import VHeader from 'components/v-header/v-header'
  import Goods from 'components/goods/goods'
  import Ratings from 'components/ratings/ratings'
  import Seller from 'components/seller/seller'
  import Tab from 'components/tab/tab'

  export default {
    data() {
      return {
        seller: {
```

```
          id: qs.parse(location.search).id
        }
      }
    },
    computed: {
      tabs() {
        return [
          {
            label: '商品',
            component: Goods,
            data: {
              seller: this.seller
            }
          },
          {
            label: '评论',
            component: Ratings,
            data: {
              seller: this.seller
            }
          },
          {
            label: '商家',
            component: Seller,
            data: {
              seller: this.seller
            }
          }
        ]
      }
    },
    created() {
      this._getSeller()
    },
    methods: {
      _getSeller() {
        getSeller({
          id: this.seller.id
        }).then((seller) => {
          this.seller = Object.assign({}, this.seller, seller)
        })
      }
    },
    components: {
      Tab,
      VHeader
    }
  }
</script>
```

```
<style lang="stylus" scoped>
  #app
    .tab-wrapper
      position: fixed
      top: 136px
      left: 0
      right: 0
      bottom: 0
</style>
```

12.3 项目组件

项目中所有页面组件都在 views 文件夹中定义，具体组件内容介绍如下。

12.3.1 头部组件

头部组件主要展示商家的基本信息，效果如图 12.4 所示。

图 12.4 头部组件效果

代码如下：

```
<template>
  <div class="header" @click="showDetail">
    <div class="content-wrapper">
      <div class="avatar">
        <img width="64" height="64" :src="seller.avatar">
      </div>
      <div class="content">
        <div class="title">
          <span class="brand"></span>
          <span class="name">{{seller.name}}</span>
        </div>
        <div class="description">
          {{seller.description}}/{{seller.deliveryTime}}分钟送达
        </div>
        <div v-if="seller.supports" class="support">

<support-ico :size=1 :type="seller.supports[0].type"></support-ico>
          <span class="text">{{seller.supports[0].description}}</span>
        </div>
```

```
        </div>
        <div v-if="seller.supports" class="support-count">
          <span class="count">{{seller.supports.length}}个</span>
          <i class="icon-keyboard_arrow_right"></i>
        </div>
      </div>
      <div class="bulletin-wrapper">
        <span class="bulletin-title"></span><span class="bulletin-text">{{seller.
bulletin}}</span>
        <i class="icon-keyboard_arrow_right"></i>
      </div>
      <div class="background">
        <img :src="seller.avatar" width="100%" height="100%">
      </div>
    </div>
</template>

<script type="text/ecmascript-6">
  import SupportIco from 'components/support-ico/support-ico'

  export default {
    name: 'v-header',
    props: {
      seller: {
        type: Object,
        default() {
          return {}
        }
      }
    },
    methods: {
      showDetail() {
        this.headerDetailComp = this.headerDetailComp || this.$createHeaderDetail({
          $props: {
            seller: 'seller'
          }
        })
        this.headerDetailComp.show()
      }
    },
    components: {
      SupportIco
    }
  }
</script>

<style lang="stylus" rel="stylesheet/stylus">
  @import "~common/stylus/mixin"
  @import "~common/stylus/variable"
```

```
.header
  position: relative
  overflow: hidden
  color: $color-white
  background: $color-background-ss
  .content-wrapper
    position: relative
    display: flex
    align-items: center
    padding: 24px 12px 18px 24px
    .avatar
      flex: 0 0 64px
      width: 64px
      margin-right: 16px
      img
        border-radius: 2px
    .content
      flex: 1
      .title
        display: flex
        align-items: center
        margin-bottom: 8px
        .brand
          width: 30px
          height: 18px
          bg-image('brand')
          background-size: 30px 18px
          background-repeat: no-repeat
        .name
          margin-left: 6px
          font-size: $fontsize-large
          font-weight: bold
      .description
        margin-bottom: 8px
        line-height: 12px
        font-size: $fontsize-small
      .support
        display: flex
        align-items: center
        .support-ico
          margin-right: 4px
        .text
          line-height: 12px
          font-size: $fontsize-small-s

    .support-count
      position: absolute
      right: 12px
      bottom: 14px
      display: flex
```

```
        align-items: center
        padding: 0 8px
        height: 24px
        line-height: 24px
        text-align: center
        border-radius: 14px
        background: $color-background-sss
        .count
          font-size: $fontsize-small-s
        .icon-keyboard_arrow_right
          margin-left: 2px
          line-height: 24px
          font-size: $fontsize-small-s

    .bulletin-wrapper
      position: relative
      display: flex
      align-items: center
      height: 28px
      line-height: 28px
      padding: 0 8px
      background: $color-background-sss
      .bulletin-title
        flex: 0 0 22px
        width: 22px
        height: 12px
        margin-right: 4px
        bg-image('bulletin')
        background-size: 22px 12px
        background-repeat: no-repeat
      .bulletin-text
        flex: 1
        white-space: nowrap
        overflow: hidden
        text-overflow: ellipsis
        font-size: $fontsize-small-s
      .icon-keyboard_arrow_right
        flex: 0 0 10px
        width: 10px
        font-size: $fontsize-small-s
    .background
      position: absolute
      top: 0
      left: 0
      width: 100%
      height: 100%
      z-index: -1
      filter: blur(10px)
</style>
```

12.3.2 商品标签栏与侧边导航组件

在商家信息下方，通过商品标签栏实现商品、评论和商家信息的切换，在商品标签中，通过侧边导航实现对商品列表的滚动和分类展示等功能。效果如图 12.5 所示。

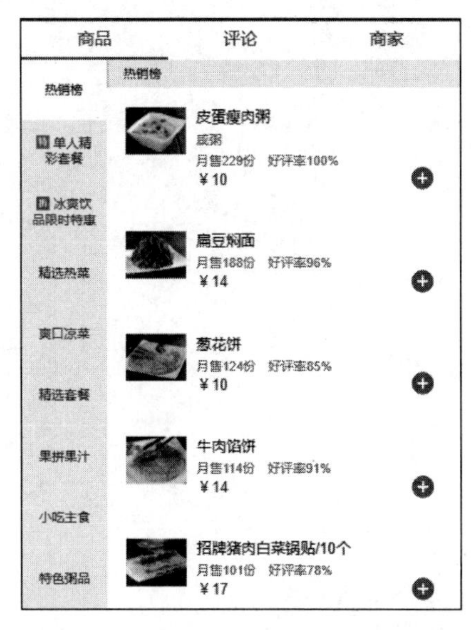

图 12.5 商品标签栏和侧边导航效果

代码如下：

```
<template>
  <div class = "tab">
    <cube - tab - bar
      :useTransition = false
      :showSlider = true
      v - model = "selectedLabel"
      :data = "tabs"
      ref = "tabBar"
      class = "border - bottom - 1px"
    >
    </cube - tab - bar>
    <div class = "slide - wrapper">
      <cube - slide
        :loop = false
        :auto - play = false
        :show - dots = false
        :initial - index = "index"
        ref = "slide"
        :options = "slideOptions"
        @scroll = "onScroll"
        @change = "onChange"
      >
```

```
        <cube-slide-item v-for="(tab, index) in tabs" :key="index">
          <component ref="component" :is="tab.component" :data="tab.data"></component>
        </cube-slide-item>
      </cube-slide>
    </div>
  </div>
</template>

<script>
  export default {
    name: 'tab',
    props: {
      tabs: {
        type: Array,
        default() {
          return []
        }
      },
      initialIndex: {
        type: Number,
        default: 0
      }
    },
    data() {
      return {
        index: this.initialIndex,
        slideOptions: {
          listenScroll: true,
          probeType: 3,
          directionLockThreshold: 0
        }
      }
    },
    computed: {
      selectedLabel: {
        get() {
          return this.tabs[this.index].label
        },
        set(newVal) {
          this.index = this.tabs.findIndex((value) => {
            return value.label === newVal
          })
        }
      }
    },
    mounted() {
      this.onChange(this.index)
    },
    methods: {
      onScroll(pos) {
```

```
        const tabBarWidth = this.$refs.tabBar.$el.clientWidth
        const slideWidth = this.$refs.slide.slide.scrollerWidth
        const transform = -pos.x / slideWidth * tabBarWidth
        this.$refs.tabBar.setSliderTransform(transform)
      },
      onChange(current) {
        this.index = current
        const instance = this.$refs.component[current]
        if (instance && instance.fetch) {
          instance.fetch()
        }
      }
    }
  }
</script>

<style lang="stylus" scoped>
  @import "~common/stylus/variable"

  .tab
    display: flex
    flex-direction: column
    height: 100%
    >>> .cube-tab
      padding: 10px 0
    .slide-wrapper
      flex: 1
      overflow: hidden
</style>
```

12.3.3 购物车组件

在没有任何商品的情况下,无法直接选择购物车组件,效果如图 12.6 所示。添加商品后,购物车将被激活,效果如图 12.7 所示。

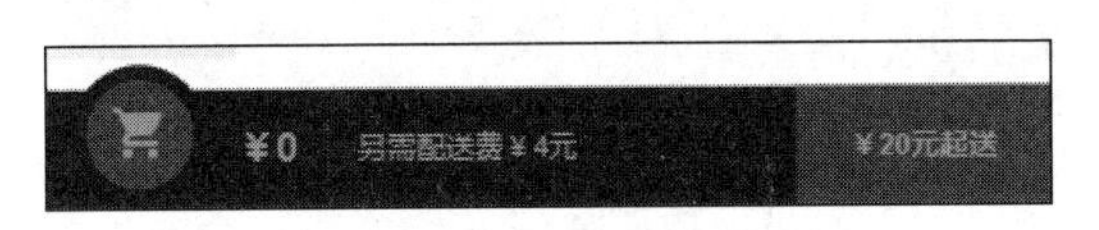

图 12.6　购物车默认状态

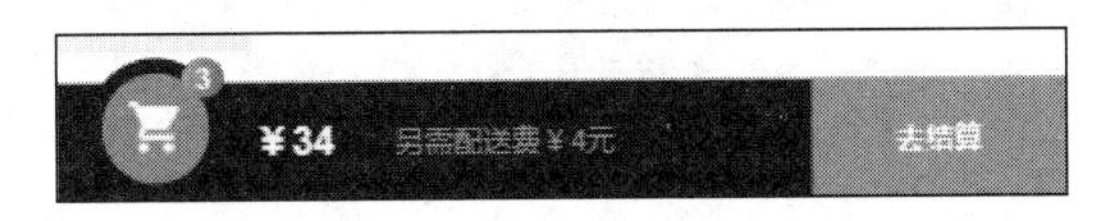

图 12.7　添加商品后的状态

单击购物车图标后,将显示用户选中的商品,效果如图 12.8 所示,在购物车商品列表页面中可以对商品进行加减操作,也可以直接清空购物车。

当单击"去结算"按钮时,将弹出购买商品花费的金额提示对话框,效果如图 12.9 所示。

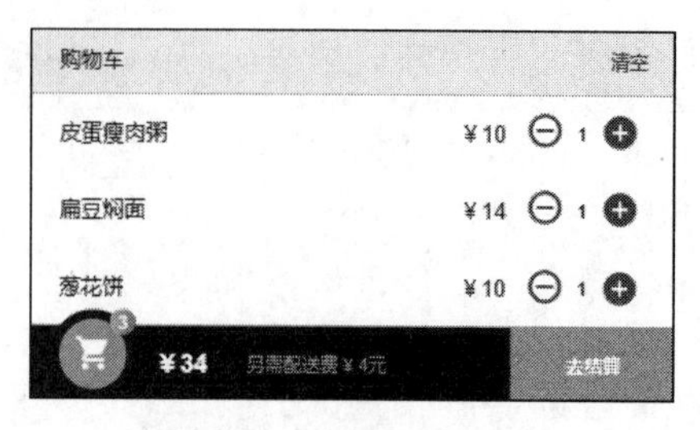

图 12.8　购物车商品列表

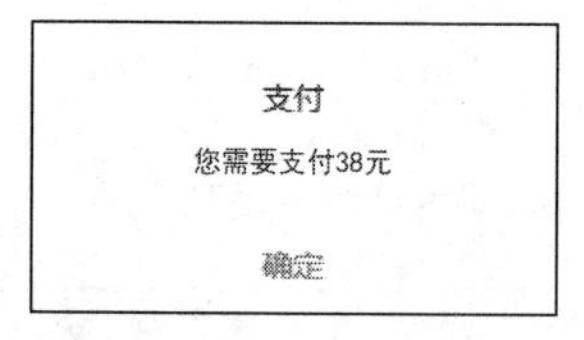

图 12.9　提示对话框

具体实现的代码存放在 shop-cart.vue 文件中。

商品购物车组件 shop-cart.vue 文件代码如下：

```
<template>
  <div>
    <div class = "shopcart">
      <div class = "content" @click = "toggleList">
        <div class = "content - left">
          <div class = "logo - wrapper">
            <div class = "logo" :class = "{ 'highlight':totalCount > 0}">
              <i class = "icon - shopping_cart" :class = "{ 'highlight':totalCount > 0}"></i>
            </div>
            <div class = "num" v - show = "totalCount > 0">
              <bubble :num = "totalCount"></bubble>
            </div>
          </div>
          <div class = "price" :class = "{ 'highlight':totalPrice > 0}">¥ {{totalPrice}}</div>
          <div class = "desc">另需配送费 ¥ {{deliveryPrice}}元</div>
        </div>
        <div class = "content - right" @click = "pay">
          <div class = "pay" :class = "payClass">
            {{payDesc}}
          </div>
        </div>
      </div>
      <div class = "ball - container">
        <div v - for = "(ball,index) in balls" :key = "index">
          <transition
            @before - enter = "beforeDrop"
            @enter = "dropping"
            @after - enter = "afterDrop">
            <div class = "ball" v - show = "ball.show">
              <div class = "inner inner - hook"></div>
            </div>
          </transition>
        </div>
      </div>
    </div>
```

```
  </div>
</template>

<script>
  import Bubble from 'components/bubble/bubble'

  const BALL_LEN = 10
  const innerClsHook = 'inner-hook'

  function createBalls() {
    let balls = []
    for (let i = 0; i < BALL_LEN; i++) {
      balls.push({show: false})
    }
    return balls
  }

  export default {
    name: 'shop-cart',
    props: {
      selectFoods: {
        type: Array,
        default() {
          return []
        }
      },
      deliveryPrice: {
        type: Number,
        default: 0
      },
      minPrice: {
        type: Number,
        default: 0
      },
      sticky: {
        type: Boolean,
        default: false
      },
      fold: {
        type: Boolean,
        default: true
      }
    },
    data() {
      return {
        balls: createBalls(),
        listFold: this.fold
      }
    },
    created() {
```

```
        this.dropBalls = []
      },
      computed: {
        totalPrice() {
          let total = 0
          this.selectFoods.forEach((food) => {
            total += food.price * food.count
          })
          return total
        },
        totalCount() {
          let count = 0
          this.selectFoods.forEach((food) => {
            count += food.count
          })
          return count
        },
        payDesc() {
          if (this.totalPrice === 0) {
            return '¥ ${this.minPrice}元起送'
          } else if (this.totalPrice < this.minPrice) {
            let diff = this.minPrice - this.totalPrice
            return '还差¥ ${diff}元起送'
          } else {
            return '去结算'
          }
        },
        payClass() {
          if (!this.totalCount || this.totalPrice < this.minPrice) {
            return 'not-enough'
          } else {
            return 'enough'
          }
        }
      },
      methods: {
        toggleList() {
          if (this.listFold) {
            if (!this.totalCount) {
              return
            }
            this.listFold = false
            this._showShopCartList()
            this._showShopCartSticky()
          } else {
            this.listFold = true
            this._hideShopCartList()
          }
        },
        pay(e) {
```

```
    if (this.totalPrice < this.minPrice) {
      return
    }
    this.$createDialog({
      title: '支付',
      content: `你需要支付${this.totalPrice}元`
    }).show()
    e.stopPropagation()
  },
  drop(el) {
    for (let i = 0; i < this.balls.length; i++) {
      const ball = this.balls[i]
      if (!ball.show) {
        ball.show = true
        ball.el = el
        this.dropBalls.push(ball)
        return
      }
    }
  },
  beforeDrop(el) {
    const ball = this.dropBalls[this.dropBalls.length - 1]
    const rect = ball.el.getBoundingClientRect()
    const x = rect.left - 32
    const y = -(window.innerHeight - rect.top - 22)
    el.style.display = ''
    el.style.transform = el.style.webkitTransform = `translate3d(0,${y}px,0)`
    const inner = el.getElementsByClassName(innerClsHook)[0]
    inner.style.transform = inner.style.webkitTransform = `translate3d(${x}px,0,0)`
  },
  dropping(el, done) {
    this._reflow = document.body.offsetHeight
    el.style.transform = el.style.webkitTransform = 'translate3d(0,0,0)'
    const inner = el.getElementsByClassName(innerClsHook)[0]
    inner.style.transform = inner.style.webkitTransform = 'translate3d(0,0,0)'
    el.addEventListener('transitionend', done)
  },
  afterDrop(el) {
    const ball = this.dropBalls.shift()
    if (ball) {
      ball.show = false
      el.style.display = 'none'
    }
  },
  _showShopCartList() {
    this.shopCartListComp = this.shopCartListComp || this.$createShopCartList({
      $props: {
        selectFoods: 'selectFoods'
      },
      $events: {
```

```
            leave: () => {
              this._hideShopCartSticky()
            },
            hide: () => {
              this.listFold = true
            },
            add: (el) => {
              this.shopCartStickyComp.drop(el)
            }
          }
        })
        this.shopCartListComp.show()
      },
      _showShopCartSticky() {
        this.shopCartStickyComp = this.shopCartStickyComp || this.$createShopCartSticky({
          $props: {
            selectFoods: 'selectFoods',
            deliveryPrice: 'deliveryPrice',
            minPrice: 'minPrice',
            fold: 'listFold',
            list: this.shopCartListComp
          }
        })
        this.shopCartStickyComp.show()
      },
      _hideShopCartList() {
        const list = this.sticky ? this.$parent.list : this.shopCartListComp
        list.hide && list.hide()
      },
      _hideShopCartSticky() {
        this.shopCartStickyComp.hide()
      }
    },
    watch: {
      fold(newVal) {
        this.listFold = newVal
      },
      totalCount(count) {
        if (!this.fold && count === 0) {
          this._hideShopCartList()
        }
      }
    },
    components: {
      Bubble
    }
  }
</script>

<style lang="stylus" scoped>
```

```
@import "~common/stylus/mixin"
@import "~common/stylus/variable"

.shopcart
  height: 100%
  .content
    display: flex
    background: $color-background
    font-size: 0
    color: $color-light-grey
    .content-left
      flex: 1
      .logo-wrapper
        display: inline-block
        vertical-align: top
        position: relative
        top: -10px
        margin: 0 12px
        padding: 6px
        width: 56px
        height: 56px
        box-sizing: border-box
        border-radius: 50%
        background: $color-background
        .logo
          width: 100%
          height: 100%
          border-radius: 50%
          text-align: center
          background: $color-dark-grey
          &.highlight
            background: $color-blue
          .icon-shopping_cart
            line-height: 44px
            font-size: $fontsize-large-xxx
            color: $color-light-grey
            &.highlight
              color: $color-white
        .num
          position: absolute
          top: 0
          right: 0
      .price
        display: inline-block
        vertical-align: top
        margin-top: 12px
        line-height: 24px
        padding-right: 12px
        box-sizing: border-box
        border-right: 1px solid rgba(255, 255, 255, 0.1)
```

```
            font - weight: 700
            font - size:  $ fontsize - large
            &.highlight
              color:  $ color - white
          .desc
            display: inline - block
            vertical - align: top
            margin: 12px 0 0 12px
            line - height: 24px
            font - size:  $ fontsize - small - s
        .content - right
          flex: 0 0 105px
          width: 105px
          .pay
            height: 48px
            line - height: 48px
            text - align: center
            font - weight: 700
            font - size:  $ fontsize - small
            &.not - enough
              background:  $ color - dark - grey
            &.enough
              background:  $ color - green
              color:  $ color - white
      .ball - container
        .ball
          position: fixed
          left: 32px
          bottom: 22px
          z - index: 200
          transition: all 0.4s cubic - bezier(0.49,  - 0.29, 0.75, 0.41)
          .inner
            width: 16px
            height: 16px
            border - radius: 50 %
            background:  $ color - blue
            transition: all 0.4s linear
</style>
```

商品购物车列表组件 shop-cart-list.vue 文件代码如下：

```
<template>
  <transition name = "fade">
    <cube - popup
      :mask - closable = true
      v - show = "visible"
      @mask - click = "maskClick"
      position = "bottom"
      type = "shop - cart - list"
```

```
        :z-index=90
      >
        <transition
          name="move"
          @after-leave="afterLeave"
        >
          <div v-show="visible">
            <div class="list-header">
              <h1 class="title">购物车</h1>
              <span class="empty" @click="empty">清空</span>
            </div>
            <cube-scroll class="list-content" ref="listContent">
              <ul>
                <li
                  class="food"
                  v-for="(food,index) in selectFoods"
                  :key="index"
                >
                  <span class="name">{{food.name}}</span>
                  <div class="price">
                    <span>¥{{food.price * food.count}}</span>
                  </div>
                  <div class="cart-control-wrapper">
                    <cart-control @add="onAdd" :food="food"></cart-control>
                  </div>
                </li>
              </ul>
            </cube-scroll>
          </div>
        </transition>
      </cube-popup>
    </transition>
</template>

<script>
  import CartControl from 'components/cart-control/cart-control'
  import popupMixin from 'common/mixins/popup'

  const EVENT_SHOW = 'show'
  const EVENT_ADD = 'add'
  const EVENT_LEAVE = 'leave'

  export default {
    name: 'shop-cart-list',
    mixins: [popupMixin],
    props: {
      selectFoods: {
        type: Array,
        default() {
          return []
```

```
        }
      }
    },
    created() {
      this.$on(EVENT_SHOW, () => {
        this.$nextTick(() => {
          this.$refs.listContent.refresh()
        })
      })
    },
    methods: {
      onAdd(target) {
        this.$emit(EVENT_ADD, target)
      },
      afterLeave() {
        this.$emit(EVENT_LEAVE)
      },
      maskClick() {
        this.hide()
      },
      empty() {
        this.dialogComp = this.$createDialog({
          type: 'confirm',
          content: '清空购物车?',
          $events: {
            confirm: () => {
              this.selectFoods.forEach((food) => {
                food.count = 0
              })
              this.hide()
            }
          }
        })
        this.dialogComp.show()
      }
    },
    components: {
      CartControl
    }
  }
</script>

<style lang="stylus" scoped>
  @import "~common/stylus/variable"
  .cube-shop-cart-list
    bottom: 48px
    &.fade-enter, &.fade-leave-active
      opacity: 0
    &.fade-enter-active, &.fade-leave-active
      transition: all .3s ease-in-out
```

```
    .move-enter, .move-leave-active
      transform: translate3d(0, 100%, 0)
    .move-enter-active, .move-leave-active
      transition: all .3s ease-in-out
    .list-header
      height: 40px
      line-height: 40px
      padding: 0 18px
      background: $color-background-ssss
      .title
        float: left
        font-size: $fontsize-medium
        color: $color-dark-grey
      .empty
        float: right
        font-size: $fontsize-small
        color: $color-blue

    .list-content
      padding: 0 18px
      max-height: 217px
      overflow: hidden
      background: $color-white
      .food
        position: relative
        padding: 12px 0
        box-sizing: border-box
        .name
          line-height: 24px
          font-size: $fontsize-medium
          color: $color-dark-grey
        .price
          position: absolute
          right: 90px
          bottom: 12px
          line-height: 24px
          font-weight: 700
          font-size: $fontsize-medium
          color: $color-red
        .cart-control-wrapper
          position: absolute
          right: 0
          bottom: 6px

</style>
```


12.3.4 商品列表组件

在商品标签页面中,商品列表主要展示所有商品的信息,可以单击商品卡片右侧的加号添加购物车。效果如图 12.10 所示。

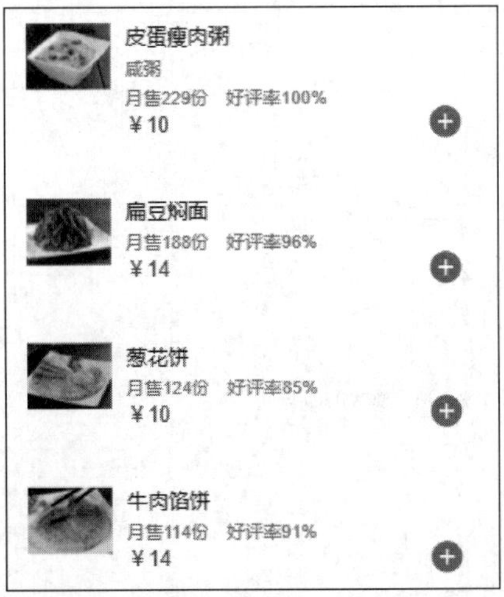

图 12.10　商品列表效果

代码如下：

```
<template>
  <div class="goods">
    <div class="scroll-nav-wrapper">
      <cube-scroll-nav
        :side=true
        :data="goods"
        :options="scrollOptions"
        v-if="goods.length"
      >
        <template slot="bar" slot-scope="props">
          <cube-scroll-nav-bar
            direction="vertical"
            :labels="props.labels"
            :txts="barTxts"
            :current="props.current"
          >
            <template slot-scope="props">
              <div class="text">
                <support-ico
                  v-if="props.txt.type>=1"
                  :size=3
                  :type="props.txt.type"
                ></support-ico>
                <span>{{props.txt.name}}</span>
                <span class="num" v-if="props.txt.count">
                  <bubble :num="props.txt.count"></bubble>
                </span>
              </div>
            </template>
```

```
        </cube-scroll-nav-bar>
      </template>
      <cube-scroll-nav-panel
        v-for="good in goods"
        :key="good.name"
        :label="good.name"
        :title="good.name"
      >
        <ul>
          <li
            @click="selectFood(food)"
            v-for="food in good.foods"
            :key="food.name"
            class="food-item"
          >
            <div class="icon">
              <img width="57" height="57" :src="food.icon">
            </div>
            <div class="content">
              <h2 class="name">{{food.name}}</h2>
              <p class="desc">{{food.description}}</p>
              <div class="extra">
                <span class="count">月售{{food.sellCount}}份</span><span>好评率
{{food.rating}}%</span>
              </div>
              <div class="price">
                <span class="now">¥{{food.price}}</span>
                <span class="old" v-show="food.oldPrice">¥{{food.oldPrice}}</span>
              </div>
              <div class="cart-control-wrapper">
                <cart-control @add="onAdd" :food="food"></cart-control>
              </div>
            </div>
          </li>
        </ul>
      </cube-scroll-nav-panel>
    </cube-scroll-nav>
  </div>
  <div class="shop-cart-wrapper">
    <shop-cart
      ref="shopCart"
      :select-foods="selectFoods"
      :delivery-price="seller.deliveryPrice"
      :min-price="seller.minPrice"></shop-cart>
  </div>
 </div>
</template>

<script>
  import { getGoods } from 'api'
```

```
import CartControl from 'components/cart-control/cart-control'
import ShopCart from 'components/shop-cart/shop-cart'
import Food from 'components/food/food'
import SupportIco from 'components/support-ico/support-ico'
import Bubble from 'components/bubble/bubble'

export default {
  name: 'goods',
  props: {
    data: {
      type: Object,
      default() {
        return {}
      }
    }
  },
  data() {
    return {
      goods: [],
      selectedFood: {},
      scrollOptions: {
        click: false,
        directionLockThreshold: 0
      }
    }
  },
  computed: {
    seller() {
      return this.data.seller
    },
    selectFoods() {
      let foods = []
      this.goods.forEach((good) => {
        good.foods.forEach((food) => {
          if (food.count) {
            foods.push(food)
          }
        })
      })
      return foods
    },
    barTxts() {
      let ret = []
      this.goods.forEach((good) => {
        const {type, name, foods} = good
        let count = 0
        foods.forEach((food) => {
          count += food.count || 0
        })
        ret.push({
```

```
          type,
          name,
          count
        })
      })
      return ret
    }
  },
  methods: {
    fetch() {
      if (!this.fetched) {
        this.fetched = true
        getGoods({
          id: this.seller.id
        }).then((goods) => {
          this.goods = goods
        })
      }
    },
    selectFood(food) {
      this.selectedFood = food
      this._showFood()
      this._showShopCartSticky()
    },
    onAdd(target) {
      this.$refs.shopCart.drop(target)
    },
    _showFood() {
      this.foodComp = this.foodComp || this.$createFood({
        $props: {
          food: 'selectedFood'
        },
        $events: {
          add: (target) => {
            this.shopCartStickyComp.drop(target)
          },
          leave: () => {
            this._hideShopCartSticky()
          }
        }
      })
      this.foodComp.show()
    },
    _showShopCartSticky() {
      this.shopCartStickyComp = this.shopCartStickyComp || this.$createShopCartSticky({
        $props: {
          selectFoods: 'selectFoods',
          deliveryPrice: this.seller.deliveryPrice,
          minPrice: this.seller.minPrice,
          fold: true
```

```
          }
        })
        this.shopCartStickyComp.show()
      },
      _hideShopCartSticky() {
        this.shopCartStickyComp.hide()
      }
    },
    components: {
      Bubble,
      SupportIco,
      CartControl,
      ShopCart,
      Food
    }
  }
</script>

<style lang="stylus" scoped>
  @import "~common/stylus/mixin"
  @import "~common/stylus/variable"
  .goods
    position: relative
    text-align: left
    height: 100%
    .scroll-nav-wrapper
      position: absolute
      width: 100%
      top: 0
      left: 0
      bottom: 48px
    >>> .cube-scroll-nav-bar
      width: 80px
      white-space: normal
      overflow: hidden
    >>> .cube-scroll-nav-bar-item
      padding: 0 10px
      display: flex
      align-items: center
      height: 56px
      line-height: 14px
      font-size: $fontsize-small
      background: $color-background-ssss
      .text
        flex: 1
        position: relative
      .num
        position: absolute
        right: -8px
        top: -10px
```

```
    .support - ico
      display: inline - block
      vertical - align: top
      margin - right: 4px
>>> .cube - scroll - nav - bar - item_active
    background: $ color - white
    color: $ color - dark - grey
>>> .cube - scroll - nav - panel - title
    padding - left: 14px
    height: 26px
    line - height: 26px
    border - left: 2px solid $ color - col - line
    font - size: $ fontsize - small
    color: $ color - grey
    background: $ color - background - ssss
  .food - item
    display: flex
    margin: 18px
    padding - bottom: 18px
    position: relative
    &:last - child
      border - none()
      margin - bottom: 0
    .icon
      flex: 0 0 57px
      margin - right: 10px
      img
        height: auto
    .content
      flex: 1
      .name
        margin: 2px 0 8px 0
        height: 14px
        line - height: 14px
        font - size: $ fontsize - medium
        color: $ color - dark - grey
      .desc, .extra
        line - height: 10px
        font - size: $ fontsize - small - s
        color: $ color - light - grey
      .desc
        line - height: 12px
        margin - bottom: 8px
      .extra
        .count
          margin - right: 12px
      .price
        font - weight: 700
        line - height: 24px
        .now
```

```
            margin-right: 8px
            font-size: $fontsize-medium
            color: $color-red
          .old
            text-decoration: line-through
            font-size: $fontsize-small-s
            color: $color-light-grey
      .cart-control-wrapper
        position: absolute
        right: 0
        bottom: 12px
    .shop-cart-wrapper
      position: absolute
      left: 0
      bottom: 0
      z-index: 50
      width: 100%
      height: 48px
</style>
```

12.3.5 商家公告组件

单击头部区域,会弹出商家公告的详细内容,效果如图 12.11 所示。

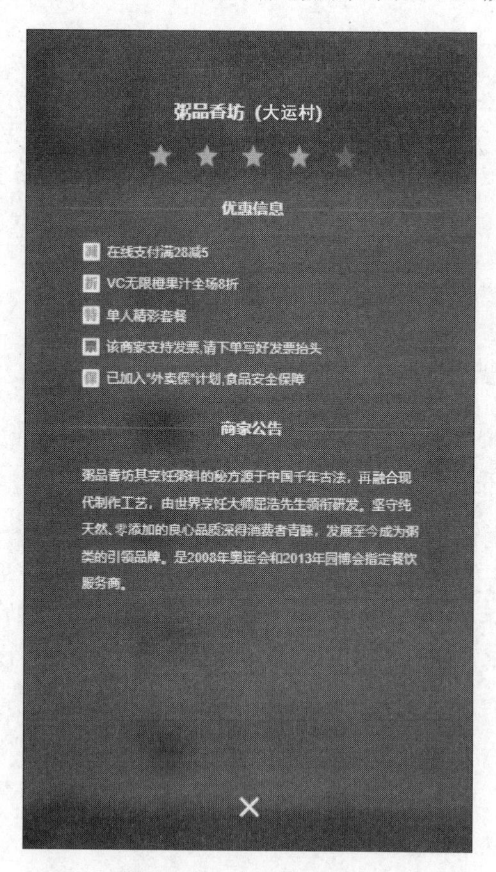

图 12.11 商家公告内容

代码如下：

```
<template>
  <transition name="fade">
    <div v-show="visible" class="header-detail" @touchmove.stop.prevent>
      <div class="detail-wrapper clear-fix">
        <div class="detail-main">
          <h1 class="name">{{seller.name}}</h1>
          <div class="star-wrapper">
            <star :size="48" :score="seller.score"></star>
          </div>
          <div class="title">
            <div class="line"></div>
            <div class="text">优惠信息</div>
            <div class="line"></div>
          </div>
          <ul v-if="seller.supports" class="supports">
            <li class="support-item" v-for="(item,index) in seller.supports" :key="item.id">
              <support-ico :size=2 :type="seller.supports[index].type"></support-ico>
              <span class="text">{{seller.supports[index].description}}</span>
            </li>
          </ul>
          <div class="title">
            <div class="line"></div>
            <div class="text">商家公告</div>
            <div class="line"></div>
          </div>
          <div class="bulletin">
            <p class="content">{{seller.bulletin}}</p>
          </div>
        </div>
      </div>
      <div class="detail-close" @click="hide">
        <i class="icon-close"></i>
      </div>
    </div>
  </transition>
</template>

<script>
  import popupMixin from 'common/mixins/popup'
  import Star from 'components/star/star'
  import SupportIco from 'components/support-ico/support-ico'

  export default {
    name: 'header-detail',
    mixins: [popupMixin],
    props: {
      seller: {
        type: Object,
```

```
        default() {
          return {}
        }
      }
    },
    components: {
      SupportIco,
      Star
    }
  }
</script>

<style lang="stylus" scoped>
  @import "~common/stylus/mixin"
  @import "~common/stylus/variable"

  .header-detail
    position: fixed
    z-index: 100
    top: 0
    left: 0
    width: 100%
    height: 100%
    overflow: auto
    backdrop-filter: blur(10px)
    opacity: 1
    color: $color-white
    background: $color-background-s
    &.fade-enter-active, &.fade-leave-active
      transition: all 0.5s
    &.fade-enter, &.fade-leave-active
      opacity: 0
      background: $color-background
    .detail-wrapper
      display: inline-block
      width: 100%
      min-height: 100%
      .detail-main
        margin-top: 64px
        padding-bottom: 64px
        .name
          line-height: 16px
          text-align: center
          font-size: $fontsize-large
          font-weight: 700
        .star-wrapper
          margin-top: 18px
          padding: 2px 0
          text-align: center
        .title
```

```
        display: flex
        width: 80%
        margin: 28px auto 24px auto
        .line
          flex: 1
          position: relative
          top: -6px
          border-bottom: 1px solid rgba(255, 255, 255, 0.2)
        .text
          padding: 0 12px
          font-weight: 700
          font-size: $fontsize-medium

      .supports
        width: 80%
        margin: 0 auto
        .support-item
          display: flex
          align-items: center
          padding: 0 12px
          margin-bottom: 12px
          &:last-child
            margin-bottom: 0
          .support-ico
            margin-right: 6px
          .text
            line-height: 16px
            font-size: $fontsize-small
      .bulletin
        width: 80%
        margin: 0 auto
        .content
          padding: 0 12px
          line-height: 24px
          font-size: $fontsize-small
    .detail-close
      position: relative
      width: 30px
      height: 30px
      margin: -64px auto 0 auto
      clear: both
      font-size: $fontsize-large-xxxx
</style>
```

12.3.6 评论内容组件

在商家评论内容的组件中，共有两个组成部分，一个是商家的评分组件，效果如图 12.12 所示；另一个是评论列表，效果如图 12.13 所示。

图 12.12 评分组件效果

图 12.13 评论列表效果

商家评分组件 ratings.vue 文件代码如下：

```
<template>
  <cube-scroll ref="scroll" class="ratings" :options="scrollOptions">
    <div class="ratings-content">
      <div class="overview">
        <div class="overview-left">
          <h1 class="score">{{seller.score}}</h1>
          <div class="title">综合评分</div>
          <div class="rank">高于周边商家{{seller.rankRate}}%</div>
        </div>
        <div class="overview-right">
          <div class="score-wrapper">
            <span class="title">服务态度</span>
            <star :size="36" :score="seller.serviceScore"></star>
            <span class="score">{{seller.serviceScore}}</span>
          </div>
          <div class="score-wrapper">
            <span class="title">商品评分</span>
            <star :size="36" :score="seller.foodScore"></star>
            <span class="score">{{seller.foodScore}}</span>
          </div>
          <div class="delivery-wrapper">
            <span class="title">送达时间</span>
            <span class="delivery">{{seller.deliveryTime}}分钟</span>
```

```
          </div>
        </div>
      </div>
      <split></split>
      <rating-select
        @select="onSelect"
        @toggle="onToggle"
        :selectType="selectType"
        :onlyContent="onlyContent"
        :ratings="ratings"
      >
      </rating-select>
      <div class="rating-wrapper">
        <ul>
          <li
            v-for="(rating, index) in computedRatings"
            :key="index"
            class="rating-item border-bottom-1px"
          >
            <div class="avatar">
              <img width="28" height="28" :src="rating.avatar">
            </div>
            <div class="content">
              <h1 class="name">{{rating.username}}</h1>
              <div class="star-wrapper">
                <star :size="24" :score="rating.score"></star>
                <span class="delivery" v-show="rating.deliveryTime">
{{rating.deliveryTime}}</span>
              </div>
              <p class="text">{{rating.text}}</p>
              <div class="recommend" v-show="rating.recommend &&
rating.recommend.length">
                <span class="icon-thumb_up"></span>
                <span
                  class="item"
                  v-for="(item, index) in rating.recommend"
                  :key="index"
                >
                  {{item}}
                </span>
              </div>
              <div class="time">
                {{format(rating.rateTime)}}
              </div>
            </div>
          </li>
        </ul>
      </div>
    </div>
  </cube-scroll>
```

```
</template>

<script>
  import Star from 'components/star/star'
  import RatingSelect from 'components/rating-select/rating-select'
  import Split from 'components/split/split'
  import ratingMixin from 'common/mixins/rating'
  import { getRatings } from 'api'
  import moment from 'moment'

  export default {
    name: 'ratings',
    mixins: [ratingMixin],
    props: {
      data: {
        type: Object
      }
    },
    data () {
      return {
        ratings: [],
        scrollOptions: {
          click: false,
          directionLockThreshold: 0
        }
      }
    },
    computed: {
      seller () {
        return this.data.seller || {}
      }
    },
    methods: {
      fetch () {
        if (!this.fetched) {
          this.fetched = true
          getRatings({
            id: this.seller.id
          }).then((ratings) => {
            this.ratings = ratings
          })
        }
      },
      format (time) {
        return moment(time).format('YYYY-MM-DD hh:mm')
      }
    },
    components: {
      Star,
      Split,
```

```
      RatingSelect
    },
    watch: {
      selectType () {
        this.$nextTick(() => {
          this.$refs.scroll.refresh()
        })
      }
    }
  }
</script>

<style lang="stylus" scoped>
  @import "~common/stylus/variable"
  @import "~common/stylus/mixin"

  .ratings
    position: relative
    text-align: left
    white-space: normal
    height: 100%
    .overview
      display: flex
      padding: 18px 0
      .overview-left
        flex: 0 0 137px
        padding: 6px 0
        width: 137px
        border-right: 1px solid $color-col-line
        text-align: center
        @media only screen and (max-width: 320px)
          flex: 0 0 120px
          width: 120px
        .score
          margin-bottom: 6px
          line-height: 28px
          font-size: $fontsize-large-xxx
          color: $color-orange
        .title
          margin-bottom: 8px
          line-height: 12px
          font-size: $fontsize-small
          color: $color-dark-grey
        .rank
          line-height: 10px
          font-size: $fontsize-small-s
          color: $color-light-grey
      .overview-right
        flex: 1
        padding: 6px 0 6px 24px
```

```
    @media only screen and (max - width: 320px)
      padding - left: 6px
    .score - wrapper
      display: flex
      align - items: center
      margin - bottom: 8px
      .title
        line - height: 18px
        font - size: $ fontsize - small
        color: $ color - dark - grey
      .star
        margin: 0 12px
      .score
        line - height: 18px
        font - size: $ fontsize - small
        color: $ color - orange
    .delivery - wrapper
      display: flex
      align - items: center
      .title
        line - height: 18px
        font - size: $ fontsize - small
        color: $ color - dark - grey
      .delivery
        margin - left: 12px
        font - size: $ fontsize - small
        color: $ color - light - grey
.rating - wrapper
  padding: 0 18px
  .rating - item
    display: flex
    padding: 18px 0
    &:last - child
      border - none()
    .avatar
      flex: 0 0 28px
      width: 28px
      margin - right: 12px
      img
        height: auto
        border - radius: 50 %
    .content
      position: relative
      flex: 1
      .name
        margin - bottom: 4px
        line - height: 12px
        font - size: $ fontsize - small - s
        color: $ color - dark - grey
      .star - wrapper
```

```
            margin-bottom: 6px
            display: flex
            align-items: center
            .star
              margin-right: 6px
            .delivery
              font-size: $fontsize-small-s
              color: $color-light-grey
          .text
            margin-bottom: 8px
            line-height: 18px
            color: $color-dark-grey
            font-size: $fontsize-small
          .recommend
            display: flex
            align-items: center
            flex-wrap: wrap
            line-height: 16px
            .icon-thumb_up, .item
              margin: 0 8px 4px 0
              font-size: $fontsize-small-s
            .icon-thumb_up
              color: $color-blue
            .item
              padding: 0 6px
              border: 1px solid $color-row-line
              border-radius: 1px
              color: $color-light-grey
              background: $color-white
          .time
            position: absolute
            top: 0
            right: 0
            line-height: 12px
            font-size: $fontsize-small
            color: $color-light-grey
</style>
```

评论列表组件 rating-select.vue 文件代码如下：

```
<template>
  <div class="rating-select">
    <div class="rating-type border-bottom-1px">
      <span @click="select(2)" class="block positive" :class="{ 'active':
selectType === 2}">{{desc.all}}<span
          class="count">{{ratings.length}}</span></span>
      <span @click="select(0)" class="block positive" :class="{ 'active':
selectType === 0}">{{desc.positive}}<span
          class="count">{{positives.length}}</span></span>
```

```
      <span @click="select(1)" class="block negative" :class="{'active':
selectType === 1}">{{desc.negative}}<span
        class="count">{{negatives.length}}</span></span>
    </div>
    <div @click="toggleContent" class="switch" :class="{'on':onlyContent}">
      <span class="icon-check_circle"></span>
      <span class="text">只看有内容的评论</span>
    </div>
  </div>
</template>
<script>
  const POSITIVE = 0
  const NEGATIVE = 1
  const ALL = 2
  const EVENT_TOGGLE = 'toggle'
  const EVENT_SELECT = 'select'

  export default {
    props: {
      ratings: {
        type: Array,
        default() {
          return []
        }
      },
      selectType: {
        type: Number,
        default: ALL
      },
      onlyContent: {
        type: Boolean,
        default: false
      },
      desc: {
        type: Object,
        default() {
          return {
            all: '全部',
            positive: '满意',
            negative: '不满意'
          }
        }
      }
    },
    computed: {
      positives() {
        return this.ratings.filter((rating) => {
          return rating.rateType === POSITIVE
        })
      },
```

```
      negatives() {
        return this.ratings.filter((rating) => {
          return rating.rateType === NEGATIVE
        })
      }
    },
    methods: {
      select(type) {
        this.$emit(EVENT_SELECT, type)
      },
      toggleContent() {
        this.$emit(EVENT_TOGGLE)
      }
    }
  }
</script>
<style lang="stylus" rel="stylesheet/stylus">
  @import "~common/stylus/variable"
  .rating-select
    .rating-type
      padding: 18px 0
      margin: 0 18px
      .block
        display: inline-block
        padding: 8px 12px
        margin-right: 8px
        line-height: 16px
        border-radius: 1px
        font-size: $fontsize-small
        color: $color-grey
        &.active
          color: $color-white
        .count
          margin-left: 2px
        &.positive
          background: $color-light-blue
          &.active
            background: $color-blue
        &.negative
          background: $color-light-grey-s
          &.active
            background: $color-grey
    .switch
      display: flex
      align-items: center
      padding: 12px 18px
      line-height: 24px
      border-bottom: 1px solid $color-row-line
      color: $color-light-grey
      &.on
```

```
        .icon-check_circle
          color: $color-green
      .icon-check_circle
        margin-right: 4px
        font-size: $fontsize-large-xxx
      .text
        font-size: $fontsize-small
</style>
```

12.3.7 商家信息组件

商家信息组件中设计了商家的星级和服务内容，效果如图 12.14 所示。

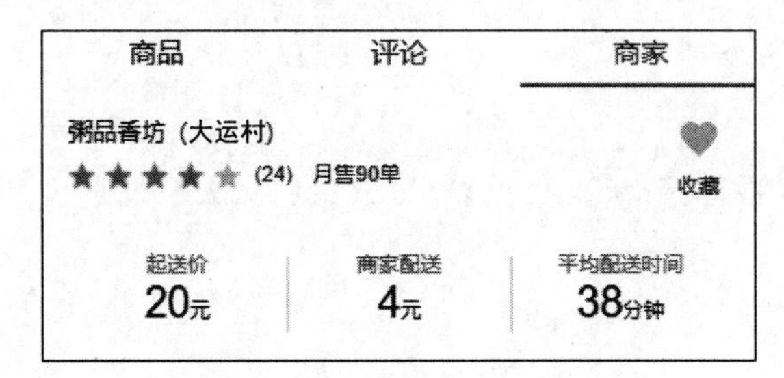

图 12.14 商家服务信息效果

商家优惠活动和公告内容效果如图 12.15 所示。

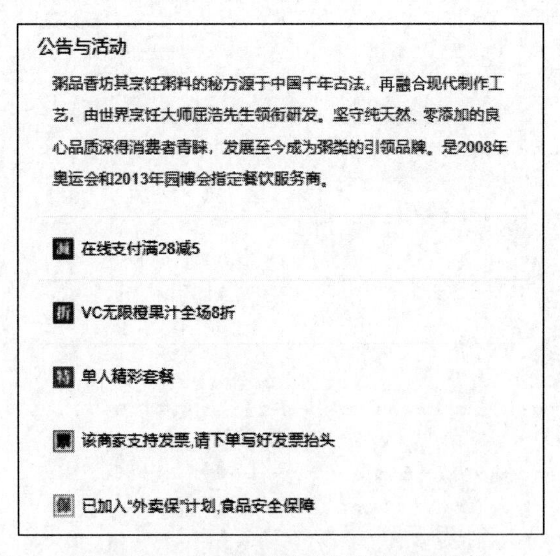

图 12.15 商家优惠活动和公告内容效果

代码如下：

```
<template>
  <cube-scroll class="seller" :options="sellerScrollOptions">
    <div class="seller-content">
```

```
<div class="overview">
  <h1 class="title">{{seller.name}}</h1>
  <div class="desc border-bottom-1px">
    <star :size="36" :score="seller.score"></star>
    <span class="text">({{seller.ratingCount}})</span>
    <span class="text">月售{{seller.sellCount}}单</span>
  </div>
  <ul class="remark">
    <li class="block">
      <h2>起送价</h2>
      <div class="content">
        <span class="stress">{{seller.minPrice}}</span>元
      </div>
    </li>
    <li class="block">
      <h2>商家配送</h2>
      <div class="content">
        <span class="stress">{{seller.deliveryPrice}}</span>元
      </div>
    </li>
    <li class="block">
      <h2>平均配送时间</h2>
      <div class="content">
        <span class="stress">{{seller.deliveryTime}}</span>分钟
      </div>
    </li>
  </ul>
  <div class="favorite" @click="toggleFavorite">
    <span class="icon-favorite" :class="{'active':favorite}"></span>
    <span class="text">{{favoriteText}}</span>
  </div>
</div>
<split></split>
<div class="bulletin">
  <h1 class="title">公告与活动</h1>
  <div class="content-wrapper border-bottom-1px">
    <p class="content">{{seller.bulletin}}</p>
  </div>
  <ul v-if="seller.supports" class="supports">
    <li
      class="support-item border-bottom-1px"
      v-for="(item,index) in seller.supports"
      :key="index"
    >
      <support-ico :size=4 :type="seller.supports[index].type"></support-ico>
      <span class="text">{{seller.supports[index].description}}</span>
    </li>
```

```
        </ul>
      </div>
      <split></split>
      <div class="pics">
        <h1 class="title">商家实景</h1>
        <cube-scroll class="pic-wrapper" :options="picScrollOptions">
          <ul class="pic-list">
            <li class="pic-item"
                v-for="(pic,index) in seller.pics"
                :key="index"
            >
              <img :src="pic" width="120" height="90">
            </li>
          </ul>
        </cube-scroll>
      </div>
      <split></split>
      <div class="info">
        <h1 class="title border-bottom-1px">商家信息</h1>
        <ul>
          <li
            class="info-item border-bottom-1px"
            v-for="(info,index) in seller.infos"
            :key="index"
          >
            {{info}}
          </li>
        </ul>
      </div>
    </div>
  </cube-scroll>
</template>

<script>
  import {saveToLocal, loadFromLocal} from 'common/js/storage'
  import Star from 'components/star/star'
  import Split from 'components/split/split'
  import SupportIco from 'components/support-ico/support-ico'

  export default {
    props: {
      data: {
        type: Object,
        default() {
          return {}
        }
      }
    },
    data() {
      return {
```

```
        favorite: false,
        sellerScrollOptions: {
          directionLockThreshold: 0,
          click: false
        },
        picScrollOptions: {
          scrollX: true,
          stopPropagation: true,
          directionLockThreshold: 0
        }
      }
    },
    computed: {
      seller() {
        return this.data.seller || {}
      },
      favoriteText() {
        return this.favorite ? '已收藏' : '收藏'
      }
    },
    created() {
      this.favorite = loadFromLocal(this.seller.id, 'favorite', false)
    },
    methods: {
      toggleFavorite() {
        this.favorite = !this.favorite
        saveToLocal(this.seller.id, 'favorite', this.favorite)
      }
    },
    components: {
      SupportIco,
      Star,
      Split
    }
  }
</script>

<style lang="stylus" scoped>
  @import "~common/stylus/variable"
  @import "~common/stylus/mixin"

  .seller
    height: 100%
    text-align: left
    .overview
      position: relative
      padding: 18px
      .title
        margin-bottom: 8px
        line-height: 14px
```

```
  font - size: $ fontsize - medium
  color: $ color - dark - grey
.desc
  display: flex
  align - items: center
  padding - bottom: 18px
  .star
    margin - right: 8px
  .text
    margin - right: 12px
    line - height: 18px
    font - size: $ fontsize - small - s
    color: $ color - grey
.remark
  display: flex
  padding - top: 18px
  .block
    flex: 1
    text - align: center
    border - right: 1px solid $ color - col - line
    &:last - child
      border: none
    h2
      margin - bottom: 4px
      line - height: 10px
      font - size: $ fontsize - small - s
      color: $ color - light - grey
    .content
      line - height: 24px
      font - size: $ fontsize - small - s
      color: $ color - dark - grey
      .stress
        font - size: $ fontsize - large - xxx
.favorite
  position: absolute
  width: 50px
  right: 11px
  top: 18px
  text - align: center
  .icon - favorite
    display: block
    margin - bottom: 4px
    line - height: 24px
    font - size: $ fontsize - large - xxx
    color: $ color - light - grey - s
    &.active
      color: $ color - red
  .text
    line - height: 10px
    font - size: $ fontsize - small - s
```

```
      color: $color-grey
.bulletin
  padding: 18px 18px 0 18px
  white-space: normal
  .title
    margin-bottom: 8px
    line-height: 14px
    color: $color-dark-grey
    font-size: $fontsize-medium
  .content-wrapper
    padding: 0 12px 16px 12px
    .content
      line-height: 24px
      font-size: $fontsize-small
      color: $color-red
  .supports
    .support-item
      display: flex
      align-items: center
      padding: 16px 12px
      &:last-child
        border-none()
    .support-ico
      margin-right: 6px
    .text
      line-height: 16px
      font-size: $fontsize-small
      color: $color-dark-grey
.pics
  padding: 18px
  .title
    margin-bottom: 12px
    line-height: 14px
    color: $color-dark-grey
    font-size: $fontsize-medium
  .pic-wrapper
    display: flex
    align-items: center
    .pic-list
      .pic-item
        display: inline-block
        margin-right: 6px
        width: 120px
        height: 90px
        &:last-child
          margin: 0
.info
  padding: 18px 18px 0 18px
  color: $color-dark-grey
  .title
```

```
            padding - bottom: 12px
            line - height: 14px
            font - size: $ fontsize - medium
          .info - item
            padding: 16px 12px
            line - height: 16px
            font - size: $ fontsize - small
            &:last - child
              border - none()
</style>
```

第13章 实战——Vue 3仿“今日头条”App

本章将开发一款仿“今日头条”的新闻App。该案例是基于Vue 3、Vue Router、Webpack和TypeScript等技术栈实现的一款新闻信息类App，适合有一定Vue框架使用经验的开发者进行学习。

13.1 项目概述

该项目是一款仿“今日头条”的新闻信息App，主要有以下功能。

(1) 新闻分类。

(2) 首页新闻列表。

(3) 刷新加载最新新闻。

(4) 用户私信留言。

(5) 新闻搜索。

(6) 查看新闻详情。

13.1.1 开发环境

本项目是基于Vue 3框架开发的一款WebApp，使用Vue CLI脚手架工具创建项目。在指定的硬盘目录处启动命令行工具，例如，在C:\\project目录下打开命令行工具，并执行以下命令：

```
#安装脚手架
npm i -g @vue/cli

#创建项目
vue create toutiao
```

参考本书11.3节的Vue 3项目创建步骤，完成“新闻头条”项目的创建。

项目创建成功后，继续在命令行工具中执行cd toutiao命令，进入项目根目录，安装Vant UI组件库和axios模块。执行命令如下：

```
#安装Vant3 UI组件库
npm i vant@next -S

#安装axios模块
npm i axios -S
```

项目的调试使用浏览器的控制台进行，在浏览器中按下 F12 键，然后单击“切换设备工具栏”，进入移动端的调试界面，可以选择相应的设备进行调试。

项目运行的效果如图 13.1 所示。

图 13.1　项目效果图

13.1.2　项目结构

项目结构如图 13.2 所示，其中 src 文件夹是项目的源文件目录，src 文件夹下的项目结构如图 13.3 所示。

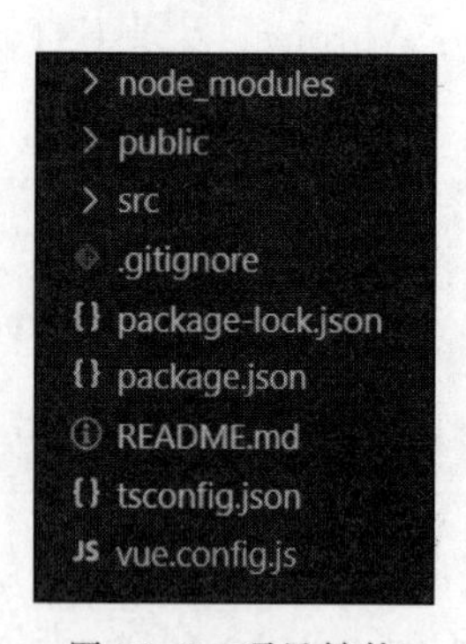

图 13.2　项目结构

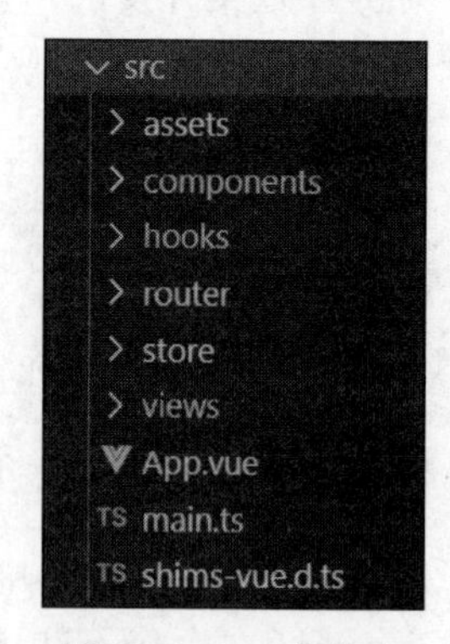

图 13.3　src 文件夹

项目结构中主要文件说明如下。

(1) node_modules：项目依赖管理目录。

(2) public：项目的静态文件存放目录，也是本地服务器的根目录。

(3) src：项目源文件存放目录。

(4) package.json：项目 npm 配置文件。

(5) vue. config. js:项目构建配置文件

src 文件夹目录说明如下。

(1) assets:静态资源文件存放目录。

(2) components:公共组件存放目录。

(3) hooks:项目的静态数据和模块封装管理目录。

(4) router:路由配置文件存放目录。

(5) store:状态管理配置存放目录。

(6) views:视图组件存放目录。

(7) App. vue:项目的根组件。

(8) main. ts:项目的入口文件。

(9) Shims-vue. d. ts: TypeScript 的适配定义文件。

13.2 入口文件

项目的入口文件有 index. html、main. ts 和 App. vue 3 个文件,这些入口文件的具体内容介绍如下。

13.2.1 项目入口页面

index. html 是项目默认的主渲染页面文件,主要用于 Vue 实例挂载点的声明与 DOM 渲染,代码如下:

```
<!DOCTYPE html>
<html lang="">
  <head>
    <meta charset="utf-8">
    <meta http-equiv="X-UA-Compatible" content="IE=edge">
    <meta name="viewport" content="width=device-width, initial-scale=1.0, maximum-scale=1.0, minimum-scale=1.0, user-scalable=0">
    <link rel="icon" href="<%= BASE_URL %>favicon.ico">
    <title><%= htmlWebpackPlugin.options.title %></title>
  </head>
  <body>
    <noscript>
      <strong>We're sorry but <%= htmlWebpackPlugin.options.title %> doesn't work properly without JavaScript enabled. Please enable it to continue.</strong>
    </noscript>
    <div id="app"></div>
    <!-- built files will be auto injected -->
  </body>
</html>
```

13.2.2 程序入口文件

main. ts 是程序的入口文件,主要用于加载各种公共组件和初始化 Vue 实例。本项目是基于 Vue 3 开发的，在入口文件中引入 createApp 来创建 Vue 实例对象。项目中的路由设置和引用的 Vant UI 组件库也是在该文件中定义的，代码如下：


```
import { createApp } from 'vue'
import App from './App.vue'
import router from './router'
import store from './store'
import Vant from 'vant';
import 'vant/lib/index.css';

const app = createApp(App)
app.use(store)
app.use(router)
app.use(Vant)
app.mount('#app')
```

13.2.3 组件入口文件

App.vue 是项目的根组件，所有的页面都是在 App.vue 下切换的，所有的页面组件都是 App.vue 的子组件。在 App.vue 组件内只需使用 <router-view> 组件作为占位符，就可以实现各个页面的引入，代码如下：

```
<template>
  <router-view/>
</template>
```

13.2.4 路由文件

在 src/router/index.ts 文件中，定义了项目的所有跳转的路由，代码如下：

```
import { createRouter, createWebHashHistory, RouteRecordRaw } from 'vue-router'

const routes: Array<RouteRecordRaw> = [
  {
    path: '/',
    name: 'Home',
    component: () => import ('@/views/Home.vue')
  },
  {
    path: '/search',
    name: 'Search',
    component: () => import ('@/views/Search.vue')
  },
  {
    path: '/details',
    name: 'Details',
```

```
    component: () => import ('@/views/Details.vue')
  },
  {
    path: '/message',
    name: 'Message',
    component: () => import ('@/views/Message.vue')
  }
]

const router = createRouter({
  history: createWebHashHistory(),
  routes
})

export default router
```

13.3 项目组件

项目中所有页面组件都在 views 文件夹中定义，所有的公共组件都在 components 文件夹中定义，具体组件内容介绍如下。

13.3.1 公共组件

在首页新闻列表和新闻内容页中，都需要使用到新闻列表的展示，为了增强代码的可扩展性，可以在 components 文件夹下创建 NewsCard.vue 新闻卡片的公共组件，代码如下：

```
<template>
  <div class="news-card" @click="onClick">
      <van-row gutter="20">
        <van-col span="16">
            <div class="news-title">{{title}}</div>
            <div class="news-msg">
                <span>{{author}}</span>
                <span>{{date}}</span>
            </div>
        </van-col>
        <van-col span="8">
            <van-image :src="pic" style="height: 100%">
                <template v-slot:error>加载失败</template>
            </van-image>
        </van-col>
      </van-row>
  </div>
</template>

<script>
export default {
```

```
    props: {
        title: String,
        author: String,
        date: String,
        pic: String
    },
    setup(props,context) {

        //新闻单击事件
        const onClick = () => {

        }

        return {
            onClick
        }
    }
}
</script>

<style scoped>
.news-card{
    box-sizing: border-box;
    padding: 0px 10px;
}
.van-row{
    padding: 15px 0px;
    border-bottom: 1px solid #F7F7F7;
}
.news-title{
    color: #222;
    font-size: 18px;
}
.news-msg{
    font-size: 12px;
    color: #999;
    margin-top: 5px;
}
.news-msg span{
    margin-right: 8px;
}
</style>
```

13.3.2 首页导航栏

App首页汇集了整个应用的核心功能入口，在头部导航栏部分，主要涉及了两个功能的入口按钮，分别是“私信”和“搜索”。在头部导航栏的下方，设计了新闻分类的标签页按钮，方便用户切换不同类别的新闻列表。效果如图13.4所示。

图 13.4　首页导航栏效果

Home.vue 首页组件，代码如下：

```
<template>
  <div class="page-body">
    <!-- 头部 -->
    <van-nav-bar fixed z-index="1000">
      <template #title>
        <span class="top-title">今日头条</span>
        <van-icon name="replay" color="#fff" size="18"/>
      </template>
      <template #left>
        <van-icon name="envelop-o" color="#fff" size="22" />
      </template>
      <template #right>
        <van-icon name="search" color="#fff" size="22" />
      </template>
    </van-nav-bar>

    <!-- 分类导航 -->
    <van-tabs
      v-model:active="active"
      background="#F4F5F6"
      sticky
      offset-top="46"
      title-active-color="#EE0A24"
    >
      <van-tab v-for="tab in tabsList"
              :key="tab.key" :title="tab.name" :name="tab.key">
        <div class="tab-views">
    <!-- 此处为新闻列表 -->
        </div>
      </van-tab>
    </van-tabs>
  </div>
</template>
```

新闻类别的标签页数据使用的是本地静态数据，在项目中的 src/hooks 文件夹下创建 tabs.ts 文件，用于保存新闻分类的静态数据，代码如下：

```
const tabs = [
    {
        key: 'top',
        name: '推荐'
    },
```

```
    {
        key: 'guonei',
        name: '国内'
    },
    {
        key: 'guoji',
        name: '国际'
    },
    {
        key: 'yule',
        name: '娱乐'
    },
    {
        key: 'tiyu',
        name: '体育'
    },
    {
        key: 'junshi',
        name: '军事'
    },
    {
        key: 'keji',
        name: '科技'
    },
    {
        key: 'caijing',
        name: '财经'
    },
    {
        key: 'shishang',
        name: '时尚'
    },
    {
        key: 'youxi',
        name: '游戏'
    },
    {
        key: 'qiche',
        name: '汽车'
    },
    {
        key: 'jiankang',
        name: '健康'
    }
]

export default tabs
```

在 Home.vue 文件中，进入 src/hooks/tabs.ts 文件，并设置为响应式数据，代码如下：

```
<script lang="ts">
import { ref } from "vue";
import tabs from "../hooks/tabs";
export default {
  name: "Home",
  setup() {
    const tabsList = ref(tabs);

    return {
      tabsList
    };
  },
};
</script>
```

13.3.3 首页新闻列表

首页新闻列表的效果如图 13.5 所示。

图 13.5 首页新闻列表的效果

在首页中引入 NewsCard.vue 新闻卡片的公共组件，并对 axios 获取的数据进行封装。由于本项目中使用的是"聚合数据平台"的数据接口，在请求服务器端接口时会出现跨域，需要先配置本地服务器的请求代理。

在项目根目录下创建 vue.config.js 配置文件，在配置文件中添加 proxy 的配置，代码如下：

```
module.exports = {
    devServer: {
      proxy: {
        '/api': {
```

```
            target: 'http://v.juhe.cn',
            changeOrigin: true,
            pathRewrite: {
              "^/api": ""
            }
          }
        }
    }
}
```

封装 axios 请求方法，在 src/hooks 目录下创建 sendhttp.ts 文件，代码如下：

```
import axios from 'axios';
import { ref } from 'vue';
import { Toast } from 'vant';

//聚合数据上申请的用户密钥
const key = 'xxx'

function sendhttp(api: string, query: string) {

    const result = ref(null)

    //加载数据
    const getData = (query: string, toast?: any) => {
        axios.get(api, {
            params: {
                type: query,
                key
            }
        }).then(res => {
            console.log(res.data)
            result.value = res.data.result.data
            if(toast){
                toast.clear()
            }
        }).catch(err =>{
            if(toast){
                toast.clear()
            }
        })
    }
    getData(query)

    //切换导航事件
    const tabsChange = (name: string) => {
        getData(name)
    }
```

```
    //刷新事件
    const replay = (name: string) => {
        const toast = Toast.loading({
            message: "加载中...",
            forbidClick: true,
            duration: 0,
            loadingType: "spinner",
        });
        getData(name, toast)
    }

    return {
        result,
        tabsChange,
        replay
    }
}

export default sendhttp
```

在 Home.vue 文件中，引入 src/hooks/sendhttp.ts 文件，并在 setup()函数中发送请求，将请求成功后的数据遍历在标签页中。

Home.vue 文件的代码如下：

```
<template>
  <div class="page-body">
    <!-- 头部 -->
    <van-nav-bar fixed z-index="1000" @click-left="onClickLeft" @click-right="
onClickRight">
      <template #title>
        <span class="top-title">新闻头条</span>
        <van-icon name="replay" color="#fff" size="18" @click="replay" />
      </template>
      <template #left>
        <van-icon name="envelop-o" color="#fff" size="22" />
      </template>
      <template #right>
        <van-icon name="search" color="#fff" size="22" />
      </template>
    </van-nav-bar>

    <!-- 分类导航 -->
    <van-tabs
      v-model:active="active"
      background="#F4F5F6"
      sticky
      offset-top="46"
      title-active-color="#EE0A24"
```

```
        @change = "tabsChange"
      >
        <van-tab v-for="tab in tabsList" :key="tab.key" :title="tab.name" :name="tab.
key">
          <div class="tab-views">
            <news-card v-for="news in result"
              :key="news.uniquekey"
              :title="news.title"
              :author="news.author_name"
              :date="news.date"
              :pic="news.thumbnail_pic_s"
              @click="onClickNews(news.uniquekey)">
            </news-card>
          </div>
        </van-tab>
      </van-tabs>
    </div>
</template>

<script lang="ts">
import { ref } from "vue";
import tabs from "../hooks/tabs";
import sendhttp from '../hooks/sendhttp';
import NewsCard from '../components/NewsCard.vue';
import { useRoute, useRouter } from 'vue-router'
export default {
  name: "Home",
  components: {
    'news-card': NewsCard
  },
  setup() {

    const router = useRouter()

    const active = ref('top');
    const tabsList = ref(tabs);

    //获取新闻数据
    const {result, tabsChange, replay} = sendhttp('/api/toutiao/index',active.value)

    //导航栏左侧按钮单击事件
    const onClickLeft = () => {
      router.push({
        path: 'message'
      })
    }

    //导航栏右侧按钮单击事件
    const onClickRight = () => {
      router.push({
```

```
      path: 'search'
    })
  }

  //单击新闻
  const onClickNews = (id:string) => {
    router.push({
      path: 'details',
      query: {
        id
      }
    })
  }

  return {
    active,
    replay,
    tabsList,
    tabsChange,
    result,
    onClickLeft,
    onClickRight,
    onClickNews
  };
 },
};
</script>

<style scoped>
.van-nav-bar {
  background-color: #d43d3d;
}
.top-title {
  font-size: 18px;
  font-weight: bold;
  color: #fff;
  margin-right: 5px;
}
.van-tabs{
  top: 46px;
}
</style>
```

13.3.4 新闻详情页

单击新闻列表，触发 NewsCard.vue 新闻卡片组件上的单击事件，通过 useRouter 路由模块，将新闻的 id 传给 Details.vue 新闻详情组件，代码如下：

```
<template>
  <div class="page-body">
    <!-- 分类导航 -->
    <van-tabs
      v-model:active="active"
      background="#F4F5F6"
      sticky
      offset-top="46"
      title-active-color="#EE0A24"
      @change="tabsChange"
    >
      <van-tab v-for="tab in tabsList" :key="tab.key" :title="tab.name" :name="tab.
key">
        <div class="tab-views">
          <news-card v-for="news in result"
            :key="news.uniquekey"
            :title="news.title"
            :author="news.author_name"
            :date="news.date"
            :pic="news.thumbnail_pic_s"
            @click="onClickNews(news.uniquekey)">
          </news-card>
        </div>
      </van-tab>
    </van-tabs>
  </div>
</template>

<script lang="ts">
import NewsCard from '../components/NewsCard.vue';
import { useRoute, useRouter } from 'vue-router'
export default {
  name: "Home",
  components: {
    'news-card': NewsCard
  },
  setup() {

    const router = useRouter()

    //其他代码

    //单击新闻
    const onClickNews = (id:string) => {
      router.push({
        path: 'details',
        query: {
          id
        }
      })
```

```
    }

    return {
      onClickNews
    };
  },
};
</script>
```

在新闻详情页中，通过 useRoute 组件获取参数，得到新闻的 id，并再次发送 axios 请求，获取当前新闻的详细数据。Details. vue 组件的代码如下：

```
<template>
  <div>
    <!-- 导航栏 -->
    <van-nav-bar left-arrow fixed @click-left="onClickLeft">
      <template #left>
        <van-icon name="arrow-left" size="16" color="#999" />
        <span style="margin-left:10px;">{{result.author_name}}</span>
      </template>
      <template #right>
        <van-icon name="share-o" size="20" color="#999" />
      </template>
    </van-nav-bar>

    <!-- 正文 -->
    <div class="news-body">
      <!-- 新闻标题 -->
      <div class="news-title">{{result.title}}</div>

      <!-- 新闻信息 -->
      <div class="news-msg">
        <span>{{result.author_name}}</span>
        <span>{{result.date}}</span>
      </div>

      <!-- 新闻内容 -->
      <div class="news-content" v-html="result.content"></div>
    </div>
  </div>
</template>

<script>
import { ref } from "vue";
import { useRoute } from "vue-router";
import axios from 'axios';

export default {
  setup() {
    const route = useRoute();
    const id = route.query.id;
```

```
    const result = ref({
        title:'',
        author_name: '',
        date: '',
        content: ''
    });

    //获取请求
    axios.get('/api/toutiao/content',{
        params: {
            uniquekey: id,
            key: '26dafe8731502872b632b9552feccf06'
        }
    }).then(res =>{
        result.value = res.data.result.detail
    }).catch(err => {

    })

    //单击导航栏左侧按钮
    const onClickLeft = () => {
      window.history.back();
    };

    return {
      onClickLeft,
      result
    };
  },
};
</script>

<style scoped>
.news-body{
    margin-top: 60px ;
    box-sizing: border-box;
    padding: 0px 15px;
}
.news-title{
    font-size: 22px;
    color: #222;
    font-weight: bold;
}
.news-msg{
    font-size: 12px;
    color: #999;
    margin: 10px 0px 20px;
}
.news-msg span{
    margin-right: 10px;
}
</style>
```

新闻详情页的效果如图 13.6 所示。

图 13.6　新闻详情页的效果

13.3.5　私信留言页

单击头部导航栏的左侧图标按钮，跳转到私信留言列表页面，效果如图 13.7 所示。

私信

独上归州 2021-3-18
你好，你的那篇文章的链接失...

椒房殿° 2021-1-8
前端简历的模板发一下吧

你在逗我笑i 2020-12-5
你好啊！

猫与玫瑰 2020-8-10
你的文章写得很棒啊！

那只小猪像你 2020-8-1
已转发

橙子姑娘 2020-5-11
已转发

忽然之间 2020-3-6
点个赞

配角戏 2020-1-3
已转发

图 13.7　私信留言列表

Message.vue私信列表组件的代码如下：

```
<template>
  <div>
      <!-- 导航栏 -->
      <van-nav-bar
        title="私信"
        fixed
        left-arrow
        @click-left="onClickLeft">
        <template #left>
            <van-icon name="arrow-left" size="16" color="#999" />
        </template>
      </van-nav-bar>

      <!-- 留言列表 -->
      <div class="message-list">
        <van-cell v-for="(msg, index) in list" :key="index" :value="msg.date">
            <!-- 使用 title 插槽来自定义标题 -->
            <template #title>
                <van-badge :dot="!msg.read">
                    <span class="custom-title">{{msg.name}}</span>
                </van-badge>
            </template>
            <template #label>
                <span class="msg-content">{{msg.content}}</span>
            </template>
        </van-cell>
      </div>

  </div>
</template>

<script>
import { ref } from 'vue';
import messageData from '../hooks/messageData';
export default {
    setup(){

        const list = ref(messageData)

        const onClickLeft = () => {
            window.history.back()
        }

        return {
            onClickLeft,
            list
        }
    }
```

```
}
</script>

<style scoped>
.message-list{
    margin-top: 50px ;
}
.msg-content{
    display: -webkit-box;
    -webkit-box-orient: vertical;
    -webkit-line-clamp: 1;
    overflow: hidden;
}
</style>
```

本项目的私信留言使用的是本地静态数据，在 src/hooks 目录下创建 messageData.ts 文件，用于留言静态数据，代码如下：

```
const messageData = [
    {
        name: '独上归州',
        date: '2021-3-18',
        content: '你好,你的那篇文章的链接失效了,能不能再重新发布一次?',
        read: false
    },
    {
        name: '椒房殿°',
        date: '2021-1-8',
        content: '前端简历的模板发一下吧',
        read: false
    },
    {
        name: '你在逗我笑 i',
        date: '2020-12-5',
        content: '你好啊!',
        read: false
    },
    {
        name: '猫与玫瑰 □',
        date: '2020-8-10',
        content: '你的文章写得很棒啊!',
        read: true
    },
    {
        name: '那只小猪像你',
        date: '2020-8-1',
        content: '已转发',
        read: true
    },
    {
```

```
        name: '橙子姑娘',
        date: '2020 - 5 - 11',
        content: '已转发',
        read: true
    },
    {
        name: '忽然之间',
        date: '2020 - 3 - 6',
        content: '点个赞',
        read: true
    },
    {
        name: '配角戏',
        date: '2020 - 1 - 3',
        content: '已转发',
        read: true
    }
]

export default messageData
```

13.3.6 新闻搜索页面

在首页的头部导航栏,单击右侧的"搜索"按钮,跳转到新闻搜索页面,效果如图 13.8 所示。

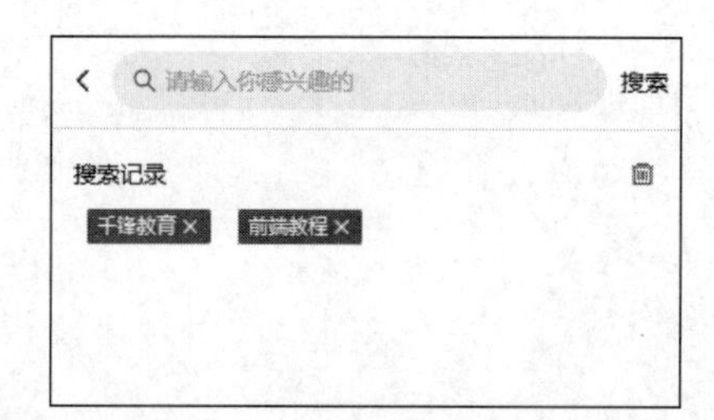

图 13.8 新闻搜索的效果

Search. vue 新闻搜索组件的代码如下:

```
<template>
  <div class = "search">
    <!-- 搜索框 -->
    <van - search
      v - model = "keywords"
      shape = "round"
      placeholder = "请输入你感兴趣的"
      show - action
      action - text = "搜索"
      @cancel = "onClickRight">
      <template #left>
```

```
        <van-icon name="arrow-left" @click="onClickLeft" style="margin-right:
10px"/>
      </template>
    </van-search>

    <!-- 搜索记录 -->
    <div class="search-history">
      <div class="search-history-title">
        <span>搜索记录</span>
        <van-icon name="delete-o" @click="clear" />
      </div>
      <div>
        <van-tag v-for="(kw,index) in list" :key="index" closeable size="medium"
type="primary" @close="delTag(index)">
          {{kw}}
        </van-tag>
      </div>
    </div>
  </div>
</template>

<script>
import { ref } from 'vue';
import { Dialog } from 'vant';
export default {
  setup() {
    const keywords = ref('')
    const list = ref(['千锋教育','前端教程'])

    //单击导航栏左侧按钮
    const onClickLeft = () => {
      window.history.back()
    }

    //单击导航栏右侧"搜索"按钮
    const onClickRight = () => {
      list.value.push(keywords.value)
      console.log(keywords)
    }

    //删除搜索记录
    const delTag = (index) => {
      list.value.splice(index,1)
    }

    //清空搜索记录
    const clear = () => {
      Dialog.confirm({
        message: '确定要清空记录吗?',
```

```
        }).then(() => {
            list.value = []
        }).catch(() => {});
    }

    return {
      keywords,
      onClickLeft,
      delTag,
      clear,
      onClickRight,
      list
    }
  }
}
</script>

<style scoped>
.van-search{
  border-bottom: 1px solid #eee;
}
.search-history{
  box-sizing: border-box;
  padding: 0px 15px;
  margin-top: 10px ;
}
.search-history-title{
  height: 30px;
  display: flex;
  align-items: center;
  justify-content: space-between;
}
.search-history-title > span{
  font-size: 14px;
}
.van-tag{
  margin: 6px 8px;
}
</style>
```